Turtles: *The Animal A...*

Turtles

The Animal Answer Guide

Whit Gibbons and Judy Greene

Color Photographs by Cris Hagen

The Johns Hopkins University Press   Baltimore

The Johns Hopkins University Press
2715 North Charles Street
Baltimore, Maryland 21218-4363
www.press.jhu.edu

Library of Congress Cataloging-in-Publication Data

Gibbons, J. Whitfield, 1939–
    Turtles : the animal answer guide / by Whit Gibbons and Judy Greene ; color
photographs by Cris Hagen.
        p. cm
    Includes bibliographical references and index.
    ISBN-13: 978-0-8018-9349-0 (hardcover : alk. paper)
    ISBN-10: 0-8018-9349-6 (hardcover : alk. paper)
    ISBN-13: 978-0-8018-9350-6 (pbk. : alk. paper)
    ISBN-10: 0-8018-9350-X (pbk. : alk. paper)
    1. Turtles. I. Greene, Judy. II. Title.
QL666.C5G53 2009
597.92—dc22   2009009806

A catalog record for this book is available from the British Library.

*Special discounts are available for bulk purchases of this book. For more information,
please contact Special Sales at 410-516-6936 or specialsales@press.jhu.edu.*

Because they are still living, turtles are common place objects to us; were they entirely extinct, their shells—the most remarkable defensive armor ever assumed by a tetrapod—would be a cause for wonder.

ALFRED SHERWOOD ROMER

Contents

Contents

Acknowledgments

We are grateful to the numerous staff and students at the Savannah River Ecology Laboratory for their encouragement and assistance in writing this book. In particular, we thank Margaret Wead for her help in preparing digital images of slides and assisting in a variety of ways with the details of preparation. We thank Kimberly Andrews, Hayley McLeod, Tom Luhring, Brian Todd, Tracey Tuberville, Tony Mills, J. D. Willson, Cris Hagen, Sean Poppy, David Scott, and Kurt Buhlmann, all of whom were helpful in making suggestions and providing insights from their own experiences with turtles. Michael Dorcas of Davidson College was also an excellent source of advice and information. Patricia West of American University provided library assistance.

We especially thank several turtle specialists for their willingness to provide comments on particular questions, although we assume full responsibility for the final answers. The following were particularly helpful: Cris Hagen, Kurt Buhlmann, Jeff Lovich, Chuck Schaffer, Tracey Tuberville, Roger Wood, Jim Knight, John Iverson, Mike Dorcas, Matt Aresco, Ken Dodd, Kim Lovich, Dick Vogt, Russ Bodie, Tony Mills, Brian Metts, J. D. Willson, Scott Pfaff, Dwight Lawson, Hal Avery, Jim Spotila, and Eugene Gaffney. We also appreciate Kurt and Cris for their numerous turtle photographs. Photographs were also contributed by David Scott, Trip Lamb, Steve Gotte, Margaret Wead, Tony Mills, Brian Metts, J. D. Willson, Hayley McLeod, Carol Gibbons, Tracey Tuberville, Mike Plummer, Justin Congdon, Tom Jones, Ken Dodd, Peri Mason, Tom Luhring, Rebecca Yeomans, and Mark Mills. The color plates in this book are from Cris Hagen's excellent collection of turtles of the world. J. D. Willson provided the cover photograph. We thank Susan Lane Harris for indexing the book.

Finally, we wish to dedicate this book to Carolyn Gibbons and Kenneth McLeod for their support and understanding during its preparation.

Introduction

In one sense this book had two beginnings, the first in July 1967 when I joined the staff of the Savannah River Ecology Laboratory (SREL) and began capturing my first turtles for research on the Savannah River Site. (I will get to the second one later.) I knew something about turtles, having captured more than 1,000 painted turtles (*Chrysemys picta*) during the previous three years while doing graduate work at Michigan State University's Kellogg Biological Station. I had even been involved in the research of Don Tinkle and Fred Cagle at Tulane University in the mid-1950s. Perhaps when I arrived at SREL I was aware that turtles would make ideal study subjects for a student interested in life history and ecology because of their traits of delaying maturity and reaching old ages that seemed distinctive from most other animals. I was keenly aware that turtles also have a variety of other traits that make them ideal for demographic studies. They are relatively easy to capture, to handle, and to mark for permanent identification. Most turtles are tough and sturdy, not fragile, making them even more ideal as study animals, especially for certain kinds of people. Among the most useful observations that can be made about individuals of many freshwater species without harming the animal are details of body size, age, sex, and state of maturity of both sexes and reproductive condition of females. Nevertheless, I still did not know enough about turtles to answer many questions about them beyond what any casual observer who spent time on southeastern rivers, lakes, and streams would know.

The second milestone was when Judy Greene came to SREL as an undergraduate research participant from Furman University in the spring of 1974 fresh from a field course in Florida on ecology and herpetology. Our formal introduction was followed by an undergraduate research and collecting trip to Kiawah Island near Charleston, South Carolina, where we captured a few slider turtles (*Trachemys scripta*). Little did we know we would catch the record body size slider turtle in the United States and that it would mark the beginning of over three decades of turtle research together. That slider size record still stands today. Since that summer, we have captured or recaptured more than 30,000 turtles of more than two dozen species from Florida to Michigan to California.

For more than 30 years, people around the world have asked us hundreds of questions about turtles, tortoises, and terrapins. Some of the commonly asked questions and some not-so-common questions follow. We

Whit Gibbons began ecological research on turtles at the Kellogg Biological Station of Michigan State University in the early 1960s. One of his first efforts was to drain a small pond in an attempt to determine whether painted turtles congregated in the mud as winter approached. Photo © Carolyn Gibbons

J. D. Willson (*left*), University of Georgia graduate student at the Savannah River Ecology Laboratory, captured an old snapping turtle. Whit Gibbons (*right*) and Judy Greene (*center*) marked the turtle before J.D. was born. SREL file photo

hope that you will find the answers to be both interesting and informative and that this book piques your interest in turtles, which, among the world's wildlife, are unique and fascinating but are all too often threatened or endangered.

Whit Gibbons

Turtles: *The Animal Answer Guide*

Chapter 1

Introducing Turtles

What are turtles?

Turtles are among the most recognizable groups of animals on earth. No one should mistake a turtle for any other type of animal. They all have a shell, four well-developed legs, and a tail, and they all lack teeth and lay shelled eggs, a combination of traits that separates them from any other kind of animal. Their distinctiveness as a natural grouping is supported by traditional paleontology and by modern-day molecular genetics—all turtles have common ancestral populations or species that lived sometime between 200 and 250 million years ago. Those early ancestors did not look very different from the turtles walking and swimming around in the world today. A signature skeletal trait of today's turtles, in addition to the shell, is the presence of a primitive skull with no opening in the temporal region on the side.

Despite their evolutionary longevity as a distinct taxonomic group the number of species alive today is relatively small. Compared with some of the other discrete vertebrate groups, such as snakes with almost 3,000 species, lizards or frogs with more than 4,000 each, and birds with as many as 10,000, the species count of modern turtles, around 320, is what scientists call "depauperate."

Turtles' distinctive suite of life history traits, including slow juvenile growth rates, attainment of reproductive maturity only after many years, and extended longevity without the signs of senility characteristic of many animals, sets them apart from many groups of animals. Turtles are ecologically and geographically diverse, occurring on all major land masses and ocean waters in warm climates and living in a diversity of freshwater, terrestrial, and marine habitats. Finally, turtles are truly an unusual biological group as they are valued by most people, feared by virtually no one, and venerated by some.

What is the difference between turtles, terrapins, and tortoises?

Oceangoing chelonians (sea turtles), the freshwater species found on every warm continent, and land-dwelling tortoises are three kinds of recognizable turtles. Thus, *turtle* is the generic name for all of the world's chelonians, including those called by other common names, such as the cooters (*Pseudemys*) of the southeastern United States and terrapins (*Malaclemys*) that inhabit brackish water along the Atlantic and Gulf coasts of North America.

Tortoises are placed in a single taxonomic family (Testudinidae) and live mostly in arid habitats, including deserts. Most tortoises have hind feet that look like the feet of a miniature elephant. The protective shell of many species is highly domed, like a World War II U.S. Army helmet. In Australia, some of the species of side-necked turtles are referred to as *tortoises*, but true tortoises in the family Testudinidae apparently never reached that continent. However, the giant and bizarre horned tortoises in the extinct family Meiolaniidae that once inhabited Australia were apparently land dwelling.

Turtles referred to as *terrapins* are found on every continent but Australia and South America and include the mangrove terrapins (*Batagur affinis* and *B. baska*) of India and Southeast Asia, painted terrapins (*Callagur borneoensis*) of Southeast Asia, and the Caspian and Spanish terrapins (*Mauremys caspica* and *M. leprosa*) of Africa, Europe, and Asia. In the United States, a species of turtle that occupies the brackish water estuaries, tidal creeks, and salt marshes of the Atlantic and Gulf coasts from Cape Cod to Texas is known as the diamondback terrapin (*Malaclemys terrapin*). In southern Florida, diamondback terrapins are found in mangrove swamps.

The word *terrapin* is more American that it might appear to some people. The Algonquin tribe of northeastern North America used "terrapin" to mean "edible turtle." These Native Americans used the name to refer to diamondback terrapins and other freshwater species they consumed for food. Because of the abundance of diamondback terrapins along the coast and because European settlers valued them for their superb culinary qualities, they acquired the name *terrapin* among English-speaking colonists. In Malaysia, the painted terrapin (known locally as "tuntung laut") inhabits the brackish waters of coastal areas, including mangrove swamps and river estuaries.

Turtles: The Animal Answer Guide

Top left, The diamondback terrapin. SREL file photo. **Top right**, impressed tortoise. Photo © Cris Hagen. **Bottom left**, river cooter. Photo © Cris Hagen. **Bottom right**, yellow-bellied slider. Photo © Trip Lamb. These are all turtles, despite their different names and habitats.

New England whalers of the nineteenth century often called Galápagos tortoises *terrapins*, or *turpin*, a term used by Charles Darwin in his journal after going ashore on one of the islands: "Met an immense Turpin; took little notice of me." The names *turpine*, *tarpain*, *turupin*, and *terapen* have also been found in the journals of early sailors referring to Galápagos tortoises. Regardless of the names given to the various species, they all qualify as *turtles*.

How many kinds of turtles are there?

Most turtle experts agree that the world has more than 300 living species of turtles that are placed in more than a dozen taxonomic families. According to our estimate, based on a combination of sources (see Appendix A), the total number of distinct species is 318 and the number of families is

General Region	Families	Species
South America	7	46
Africa	6	51
Central America (includes Caribbean)	6	45
Asia (includes Indonesia, Philippines)	6	95
North America (U.S. and Canada)	5	48
Australia (includes New Guinea)	3	37
Europe	3	6

General estimates of the numbers of turtle families and species in various regions of the world are based on cumulative totals from a variety of sources. Because of the numerous changes in turtle taxonomy proposed by turtle biologists in the early twenty-first century, the actual number of species and even families vary among sources. Some species occur in more than one general continental region and may be included more than once. The chart also does not give a clear view of how species diversity can be concentrated in one part of a continent. For example, few turtle species are found in the western three-fourths of the United States, yet more than half of the continent's species are concentrated in the southeastern portion of the country.

14. Different authorities vary in their opinions and interpretations of the exact numbers of each, although their decisions are based on the same data, as is true in many diverse taxonomic groups. Some "lumpers" among turtle taxonomists consider the number of species to be lower than that reported by "splitters." One reason is disagreement among turtle biologists about whether certain taxa should be recognized as subspecies or should be elevated to distinct species, which would create more named species.

The outcome of genetic, morphological, and paleontological information still forthcoming will help determine the relationships among some species and subspecies of turtles, but interpretations of their significance, again using the same data, will vary among turtle biologists. Nomenclatural recognition of the putative distinctiveness of questionable "species" can often be addressed by recognizing them as subspecies, but whether particular taxa receive the status of "species" is important to some systematists. One problem with providing names in perpetuity for turtles and many other groups of plants and animals is that new biological findings are constantly revising our understanding of phylogenetic relationships among species. Thus, frequent taxonomic changes are made to reflect these relationships.

One of the most obvious cases of splitting by taxonomists is that of the slider turtles, formerly recognized by most turtle biologists as a single species (*Trachemys scripta*) as recently as the early 1990s. In his excellent chapter in *Life History and Ecology of the Slider Turtle* on the distribution of the species in Mesoamerica, John Legler of the University of Utah recognized 16 subspecies within the single species. During the ensuing decade or so, 11 of the 16 were elevated to species level and only three subspecies (*T. s. scripta, T. s. elegans, and T. s. troostii*) were recognized as being the same species as the historically known slider turtle. Thus, what was taught in herpetology classes in the twentieth century as one species became recognized within the first decade of the twenty-first century as eleven distinct species. A perception by someone unfamiliar with the mercurial nature of taxonomy is that the species diversity of turtles of the world is increasing, which is absolutely not true. Obviously, the name given to an animal by humans does not change its biology or its relationship to other species. The important aspect, from the standpoint of understanding the ecology of an organism and the conservation issues associated with it, is that people know which animal is being referred to when a name is used. Taxonomists will continue to enjoy the dynamic process of working out the details while the turtles already know who they are and really do not care one way or the other what we call them.

Some of the families of turtles consist of many species whereas others have no more than a handful, and four have only one. As with species numbers, taxonomists disagree about how many families of turtles there are. The largest and most familiar groups of turtles are the hard-shelled freshwater species belonging to the families Emydidae and Geoemydidae that inhabit the Americas, Eurasia, and Africa and collectively include more than 100 species. Approximately 75 species belong to the side-necked turtle families (Chelidae, Pelomedusidae, and according to some authorities, Podocnemidae), which are primarily restricted to the Southern Hemisphere. Taxonomists are consistent in their placement of the approximately 50 species of the world's tortoises into the family Testudinidae and the 30 or so species of softshell turtles in the family Trionychidae. The family of mud turtles (Kinosternidae) has about two dozen species and ranges from Canada to Peru.

Only four species of freshwater turtles, including the giant alligator snapping turtle (*Macrochelys temminckii*) of the United States, belong to the family Chelydridae. Six species of hard-shelled sea turtles are in the family Cheloniidae. Four turtle families have only a single species each—Platysternidae, Carettochelyidae, Dermatemydidae, and the once-diverse Dermochelyidae, whose sole living representative is the world's largest and most widespread living turtle, the leatherback sea turtle (*Dermochelys coriacea*).

Why are turtles important?

Turtles are important for many reasons, including having ecological roles as major components of the food web, both as predators and prey, as well as being efficient scavengers, in many natural and human-affected ecosystems. Some species, such as the gopher tortoises (*Gopherus polyphemus*), even influence the environmental character of their habitat by digging long, deep underground burrows that are used as homes and refuges by dozens of other animals as well as by the tortoises. Turtles in some habitats, especially large rivers, lakes, and oceans, can serve as sentinels of environmental quality, based on scientific assessments of their health, regional abundance, and overall biodiversity. Finally, turtles are a much-appreciated component of human culture as part of the natural wildlife heritage of many regions, making cherished pets in some situations, and historically even being used for food, a tradition that unfortunately continues into the twenty-first century when unsustainable commercial collecting has gone unchecked.

Why should people care about turtles?

Turtles comprise a group of unique animals that are known internationally as recognizable inhabitants of land, fresh waters, and the ocean. They are part of the natural heritage of most habitable regions of the world. Therefore, turtles share the earth with us and have as much right to exist as any other animal. Turtles differ from many species that gain the attention of humans in that none prey on humans, and no species of turtle competes with humans in any appreciable way for food or other resources. Some people believe that turtles can negatively affect game fish populations by catching and eating them, but no substantive evidence has ever been produced to confirm such an idea. Indeed, the role of turtles as scavengers that remove dead fish and consumers that eat aquatic vegetation more likely enhance fishing in lakes and ponds. Although some people may perceive minimal value in turtles, others admire their symbolic traits, such as persistence, patience, and resilience. We should care about turtles because they are a unique and admirable component of native wildlife.

Where do turtles live?

Turtles live on every continent (except Antarctica) and on larger islands such as Madagascar, New Guinea, and Borneo. One or more species are often found inhabiting smaller islands, including giant tortoises on some of the Galápagos and Seychelles archipelagos, and sea turtles frequently nest on small islands. Native turtles are absent from cold temperate zone islands

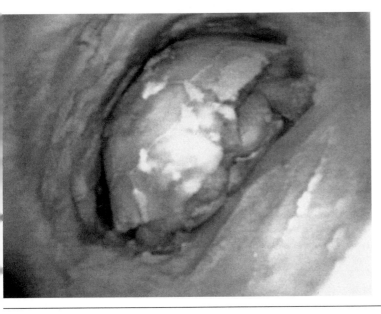

A camera reveals a gopher tortoise more than 15 feet deep in its underground burrow. The burrow camera is an infrared fiber optics cable that can be inserted into a tunnel while observers on the surface view the screen. Photo © Tracey Tuberville

such as Greenland, Baffin Island, and Great Britain. The greatest concentrations of turtles are in the warm and humid temperate zones of southeastern United States and southeastern Asia.

Freshwater turtles may live in almost any aquatic or semiaquatic habitat, including seasonal wetlands, permanent lakes, marshes, swamps, and rivers. Marine turtles collectively occupy all of the world's oceans, including occasional visits to the Arctic Ocean. Diamondback terrapins (*Malaclemys terrapin*) inhabit the narrow strip of brackish water habitat between inland fresh waters and open ocean salt water in the eastern United States, and a population of terrapins, possibly introduced, lives in brackish lakes on Bermuda. Terrestrial tortoises are native to all of the warm continents, except Australia, and to many larger islands.

What is the current classification of turtles?

The classification of turtles varies depending on which turtle biologists are asked. For the past century, turtles have been accepted by most biologists as a distinct line of vertebrates distinguishable both morphologically and phylogenetically (i.e., because of their evolutionary lineages) from all other major groups of animals. Earlier classifications were based almost solely on structure and morphology of the skull, shell, and appendages. Turtles have historically been classified as reptiles, along with crocodilians, snakes, and lizards. DNA analyses during the late twentieth century and early twenty-first century supported their recognition as a distinct line of vertebrate animals. Most authorities accept that the ancestors of birds and crocodilians were more closely related to one another than to snakes

Trammel nets set across a creek are one of the techniques used to catch some turtles, such as diamondback terrapins, that live in clean open waters without extensive vegetation or debris. The terrapins, along with crabs, stingrays, and sharks, become entangled in the three layers of fine mesh that are checked at intervals of about two hours. SREL file photo

Many species of turtles are restricted to specific habitats as with diamondback terrapins that inhabit coastal mangrove swamps of southern Florida. Photo © Kurt Buhlmann

and lizards, and that turtles had a separate origin from any of these major groups. Some biologists even take a position that turtles and crocodilians should no longer be classified as reptiles. Nonetheless, maintaining the traditional terminology that refers to turtles (as well as crocodilians) as reptiles is more efficient and less confusing because most people and most herpetological textbooks, field guides, and popular works continue to classify turtles as reptiles.

What characterizes the major groups of turtles?

Turtles are separated into two major groups (Pleurodira and Cryptodira) based on a variety of skeletal characteristics that date back to the earliest fossils. Modern turtles are readily distinguished by a simple trait: the manner in which they retract their neck. The pleurodiran turtles, also known as side-necked turtles, withdraw their neck on one side or the other beneath the carapace and plastron, with the head still visible. The cryptodiran turtles, also called the hidden-necked turtles, retract the neck and head horizontally into the front part of the body cavity, with many species able to hide the head completely by folding the shell in a protective manner. Pleurodiran turtles are represented by about 85 species of modern turtles and cryptodirans by about 230 species. Pleurodiran turtles are separated into three separate family groups (see Appendix A) based on shell characteristics. Cryptodires are divided into 11 separate families, also based on shell characteristics, but also on the presence of flippers in some (sea turtles and pig-nosed turtles), leathery shells (softshell turtles, leatherback sea turtles [*Dermochelys coriacea*], and pig-nosed turtles), or subtle skull and shell characteristics and limb structure (tortoises, freshwater turtles, and other groups). Most turtle biologists do not contest the differentiation of the major groups of turtles other than some consider the big-headed turtle of Asia (*Platysternon megacephalum*) to be in the same family as the snapping turtles (*Chelydra*) and alligator snapping turtles (*M. temminckii*). Also, some authorities support the recognition of softshell turtles as belonging to a separate suborder (Trionychoidea) than the hard-shelled cryptodires, from which they differ in many ways.

When did turtles first evolve?

The earliest turtle fossils, with shells and skeletal structures similar to modern turtles, have been identified in geological strata more than 200 million years old. The heavy bony structure of turtles has resulted in the discovery of many fossils associated with sedimentary rocks whose geological age has been determined with some certainty. The differentiation of

Pleurodires and Cryptodires is known to have occurred by the late Triassic, more than 210 million years ago, based on fossils of each group.

What is the oldest fossil turtle?

The oldest well-known fossil turtle for which nearly complete body parts are available is *Proganochelys quenstedti*, described in 1887 from a late Triassic site in Germany. The estimated age of the fossil is 210 million years. Eugene Gaffney of the American Museum of Natural History (AMNH) described the complete osteology of the species, including the shell, skull, vertebrae, pectoral and pelvic girdles, and limbs, on the basis of parts of seven different fossil specimens. The species was armored with a heavy shell, a long spiked tail, and a spiny neck. The shell of the largest *Proganochelys* specimen was more than 3 feet long, slightly larger than the largest adult alligator snapping turtle (*M. temminckii*). Another ancient fossil discussed by Gaffney, Haiyan Tong (of the AMNH), and Peter Meylan of Eckerd College is *Proterochersis robusta*, a pleurodiran turtle, also from the late Triassic of Germany. Gaffney also reported on a skull of another species of turtle from the early Jurassic that is the earliest-known turtle from the African continent. In 2008, a fossil discovered in China and called *Odontochelys semitestacea* was estimated to be from 220 million years ago. It had a partially developed carapace and is believed to be an ancestor of today's turtles.

What is the largest fossil turtle?

Determining the accurate size of the world's largest fossil turtle is difficult because few complete skeletons of the fossil giants have been recovered. Fossils often include only portions of the carapace; therefore, estimates of the actual shell size are often based on extrapolations from plates or bony parts.

The largest fossil freshwater turtle discovered (and probably the largest complete shell ever found of any type of fossil turtle) is *Stupendemys geographicus*, a side-necked turtle from Venezuela discovered and described by Roger Wood of Richard Stockton College of New Jersey. When it was originally described in 1976, the largest-known shell of *Stupendemys* measured 2.3 meters (about 7.5 feet) in length.

The full carapace has been estimated to have been even longer in life. Fossils of the genus, which is in the family Podocnemidae, have been found in Miocene-Pliocene sediments from northern Venezuela and the Amazon River Basin. When asked how he found the first and, at the time, largest specimen, Roger said he and a couple of students were in Venezuela taking a lunch break (he was eating a peanut butter and jelly sandwich) when another student walked up carrying what looked like a bone-colored briefcase under his arm. The student wasn't sure what he had found, but Roger told him that he had found the pygal bone (the last one at the end of the carapace) of what would probably prove to be the largest turtle in the world. Returning to the site where the bone had been found, Roger and his students spent two days removing the sediments covering the entire carapace. The specimen now resides in the Museo de Ciencias Naturales (MCNG), Caracas, Venezuela. Three specimens of *Stupendemys* have been discovered, and the most recently discovered shell reported to be in the MCNG may be even larger than that found by Wood.

The largest extinct marine turtles belonged to the family Protostegidae. The largest or one of the largest individuals ever recorded was a fossil specimen from South Dakota of *Archelon ischyros*, which lived in Cretaceous seas 70 million years ago. According to the description given by George Weiland in 1896, the bony shell alone of this giant sea turtle was close to 6 feet long, and the distance across the body between the tips of the two front flippers was more than 16 feet. Its head was more than 3 feet long. Considering that only the bony part of the shell was included and not the soft parts of the carapace, the total shell length may have been close to 7 or 8 feet. With the head and neck extended, the turtle was probably more than 10 or 11 feet long. Another slightly smaller sea turtle, belonging to the same family, was given the species name *Protostega gigas*. The enormous leatherback sea turtles (*Dermochelys coriacea*) of modern times are the closest living relatives to the extinct family. Several extinct species of leatherback sea turtles have already been described, many by Roger Wood, and other fossilized specimens are still being discovered. In 2007, Jim Knight of the South Carolina State Museum found fossils of an undescribed extinct species of leatherback near Summerville, South Carolina. The oceangoing giant had a carapace length estimated to be more than 7 feet long.

One of the largest-known terrestrial turtles was the Atlas tortoise, *Colossochelys atlas* (also placed by some authorities in the genus *Geochelone* or *Testudo*) of India and Pakistan. This tortoise's estimated shell length is near 7 feet and dome height is about 6 feet, although some turtle authorities believe this is an exaggeration. These enormous Pleistocene animals easily weighed a ton or more and probably represent the heaviest turtles known, extinct or living. Another large tortoise, *Meiolania*, from Lord Howe Island and the mainland of Australia was impressive because of its enormous head armed with a pair of distinctive hornlike projections that made the head in the Queensland species almost 2 feet wide.

Chapter 2

Form and Function

What are the largest and smallest living turtles?

Although only one of the living species mentioned next is indisputably the largest turtle in the world, several marine, freshwater, and terrestrial species from different regions of the planet qualify for the Big Turtle Hall of Fame. Some turtle species are clearly larger than others, but when it comes to the really big ones, confirming the record size for length or weight is difficult because people often estimate and, therefore, exaggerate size instead of using a ruler or scales. Because everybody likes to have found a record and because true measurements have sometimes been difficult to obtain, several different species have been declared the largest freshwater turtle in the world.

Declaring any species of freshwater turtle as indisputably the largest is especially problematic because of the great range in adult body sizes within many of the notably large species. Also, deciding whether maximum size of a species should be based on its length or its weight is debatable. Peter Pritchard, founder of the Chelonian Research Institute, points out that measurement of the maximum size of softshell turtles can be especially difficult because of the difference in overall length of the leathery shell compared with the "bony disc length" that measures the body length defined by the vertebrae, which he considers a better index of size.

Despite arguments about the most appropriate representation of body size, no one challenges the fact that North America's largest nonmarine turtle is the alligator snapper (*Macrochelys temminckii*) and that South America's largest turtle is the arrau river turtle (*Podocnemis expansa*). Extensive scientific information has been gathered on alligator snappers, beginning

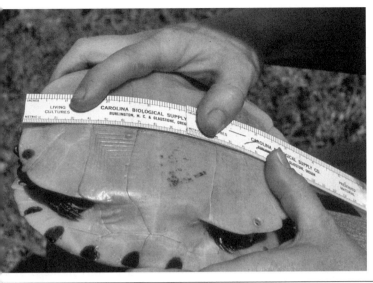

One of the most straightforward, easy, and accurate measurements of turtle size is the curved or straight-line plastron or carapace length. A flexible plastic ruler, measuring tape, or calipers have all been used historically in the field as well as in the laboratory. Photo © Whit Gibbons

as early as the 1970s by Jim Dobie of Tulane University and later Auburn University. Adult male alligator snappers get appreciably larger than females (a trait referred to as sexual dimorphism in body size), commonly reach carapace lengths of 2 feet, and weigh close to 100 pounds. Shell-length records of about 2.5 feet have been verified. According to Pritchard, author of *The Alligator Snapping Turtle*, the heaviest legitimate weight records are more than 200 pounds.

Several investigators in South America have conducted ecological studies on the arrau and have provided reliable data on body sizes. The adult female arrau, which reaches a larger size than males (the reverse form of sexual dimorphism than that seen in alligator snappers), has an average carapace length of about 26 inches (67 cm) based on several studies, and a maximum length of almost 3 feet (89 cm) has been given in several accounts, including by Peter Pritchard and Pedro Trebbau in *The Turtles of Venezuela* (1984). Dick Vogt indicates in *Amazon Turtles* (2008) that "the largest female on record measured 109 cm," which is almost 43 inches. The average body weight of these river giants is around 55 pounds (25 kg) with the maximum reported to be 160 pounds (73 kg) by Pritchard and Trebbau and 198 pounds (90 kg) by Vogt.

Dick Vogt of Instituto Nacional de Pesquisas da Amazonia (INPA), whose field studies on these and a diversity of other tropical American turtle species makes him a world expert, confirms the accuracy of the size records for the arrau. Thus, the documented records provide evidence that the South American river turtles grow to greater body lengths than alligator snappers, but the latter species is heavier. Although these are both enormous creatures, neither is truly the largest freshwater turtle in the world, according to most scientists who study turtles.

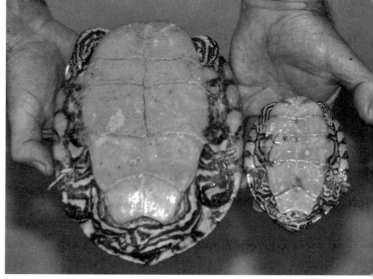

Sexual dimorphism is expressed in different ways among turtle species. *Top*, In some aquatic species such as slider turtles, males have elongated foreclaws that are used in courtship. Photo © Whit Gibbons. ***Bottom***, The size difference between an adult female Barbour's map turtle (*Graptemys barbouri*) and the much smaller male is impressive. Photo © Cris Hagen

Few ecological studies on large freshwater turtles of Asia have been as extensive as those in the Western Hemisphere, but on the basis of assertions in various books about individual turtles of great size, some of the species in Asia may reach longer lengths and greater weights than species on other continents. Among hard-shelled species, the Malaysian giant turtle (*Orlitia borneensis*) reaches lengths of 2.5 feet (76 cm) and weights of more than 100 pounds (45 kg).

Some softshell turtles apparently reach even larger sizes. Pritchard's list of the largest authentic records of the "giant species" of softshells is one with individuals ranging from about three up to four feet in leathery shell maximum length. Additional questionable records exist for some of the species. Included among them is a sure candidate for a horror movie— Bibron's frog-faced giant softshell turtle (*Pelochelys bibroni*), which is also known as the New Guinea giant softshell—at 3.5 feet. One report has a shell length of 5.5 feet and 550 pounds, but this measurement has not been confirmed. Southeast Asian narrow-headed softshell turtles (*Chitra chitra*) are stated as reaching carapace lengths of almost 4 feet, and Pritchard cites a 1914 report of a 6-foot-long specimen but questions its validity. The Shanghai softshell turtle (*Rafetus swinhoei*), a species on the verge of extinction, is considered by some to be the largest of all freshwater turtles. Adult females, which are larger than males, are typically 2.5 to 4 feet long, with one report of an animal nearly 6 feet long, but some turtle biologists lack confidence in the veracity of the record. The record weight reported for the species is more than 400 pounds.

The largest living land turtles are the Galápagos giant tortoises (*Chelonoidis nigra*) that live on the Galápagos Islands of Ecuador in the Pacific and the giant Aldabra tortoises (*Aldabrachelys dussumieri*) on Aldabra Island of the Seychelles in the Indian Ocean. Adult males reach straight-line shell lengths of around (or sometimes even more than) 4 feet (1.2 m). Because of their thick, high-domed shells, they are quite heavy, with captive individuals weighing from 500 to 900 pounds or more (225 to 409 kg or more), which makes weighing them a challenge.

The largest living turtle of all, both in length and in body mass, is incontestably the leatherback sea turtle (*Dermochelys coriacea*). Average-sized adults of these giants commonly exceed both the length and weight of other sea turtles and the largest freshwater turtles and tortoises. No other living species of turtle comes close to the size of the largest leatherbacks, which are more than 6 feet in shell length. The record weight reported for a leatherback sea turtle was a male that washed ashore on Harlech Beach, Wales, in St. George's Channel between the Irish Sea and North Atlantic. The animal weighed more than a ton (2,016 pounds, or 916 kg). Having a specimen of the world's heaviest living species of reptiles for which the massive dimensions could be measured was fortunate. However, the suspected cause of death was disappointing, even tragic, to turtle lovers worldwide. A necropsy revealed that the unfortunate animal apparently died from a blockage in the small intestine after the leatherback swallowed a large piece of hard plastic that it may have mistaken for a jellyfish or other prey item. Sadly, ingesting human trash, perhaps intentionally or thoughtlessly discarded, got him into the record books.

Determining the smallest turtle species is more difficult than confirming that the leatherback is the largest turtle in the world. The smallest sea turtles are Kemp's ridley (*Lepidochelys kempii*) of the Atlantic and Gulf coasts of the United States and the olive ridley (*Lepidochelys olivacea*) of the Pacific and Indian oceans. The largest adults are slightly over 2 feet long but do not reach a length of more than 2.5 feet, but they are still much larger than most species of tortoises and freshwater turtles. The bog turtle is generally considered to be the smallest species in North America because the maximum length is less than 4.5 inches. Most adults are less than 4 inches long. The females of all other U.S. turtles reach larger sizes than any bog turtles (*Glyptemys muhlenbergii*). However, the issue is confounded because adult males of some of the map turtles, such as the Texas map turtle (*Graptemys versa*) and Cagle's map turtle (*Graptemys caglei*), are smaller than either sex of adult bog turtles. These tiny male map turtles seldom reach 4 inches. Some populations of common musk turtles (*Sternotherus odoratus*) in Florida and South Carolina also have been reported to have adult males that are smaller than bog turtles. But these populations of small musk turtles are presumed to be the same species as the common musk turtle in which individuals of both sexes in most other populations grow larger than bog turtles.

Despite the small size of even the largest bog turtles, the record for the smallest turtle in the world in body length is probably the speckled padloper tortoise (*Homopus signatus*) of South Africa. The species holds the uncontested record as the smallest tortoise in the world. The maximum length of females, which are slightly larger than males, is less than 4 inches, which is smaller than adult bog turtles.

What is the metabolism of a turtle?

Turtles are ectothermic (cold-blooded) animals whose body temperature depends on the surrounding environment, in contrast to most mammals and birds that maintain a relatively constant body temperature with heat generated internally. Thus, the body temperature of a painted turtle (*Chrysemys picta*) from a cold northern region may vary during its lifetime from slightly below freezing in winter to above 90 degrees in summer. The temperature of the water, air, or soil surrounding a turtle is a major determinant of the turtle's body temperature, and metabolism increases as the turtle's internal temperature increases. However, turtles are not completely at the mercy of their external environment. Individual turtles not only can select a site based on temperature but also can influence their body temperature and metabolism through basking behavior. Turtles are much more likely to sit atop logs on sunny, spring days when air and water temperatures are cool than later in the summer when the water and air temperatures are already high. By basking, the turtle is able to raise its temperature as much as 20 or 30 degrees above the surrounding air and water, with a resulting increase in its metabolism. Turtles in low-elevation tropical regions, where warm air and water temperatures do not fluctuate appreciably, occasionally bask to raise their body temperatures, but optimal temperature levels can often be maintained without ever leaving the water. Species such as the South American toad-headed turtles (genus *Phrynops*) in high elevations in the tropics, which can have temperatures similar to the temperate zones, commonly bask.

For many years, a mystery to turtle biologists was how the giant leatherback sea turtles (*D. coriacea*) manage to keep their metabolism high enough to operate in cold Arctic waters. In addressing conservation issues for leatherbacks around the world, Jim Spotila of Drexel University investigated many aspects of leatherbacks' life history and ecology, including their thermal biology and metabolism. Spotila collaborated with several colleagues, including Ed Standora of Buffalo State University, Frank Paladino of Purdue University (Fort Wayne), and Mike O'Connor also from Drexel, to provide scientific evidence that leatherbacks are able to maintain their internal body temperature above their colder surroundings. Among the documented facts was a case study in which a leatherback maintained a body temperature of 78 degrees F (25.5 degrees C) while swimming in ocean waters of around 45 degrees F (7.5 degrees C).

Leatherbacks are able to maintain higher body temperatures using gigantothermy, more effective insulation than other turtles, and changes in their circulatory system. Gigantothermy is based on the simple principle that larger animals have a greater volume-to-surface ratio so that they lose heat to their surroundings more slowly than smaller animals of the same shape. Insulation (in the form of a fat layer) also helps maintain body temperatures, and a circulatory system designed to shunt blood to different parts of the body provides more effective warming or cooling, allowing the leatherback to adjust its body temperature internally, somewhat similarly to homeotherms, such as birds and mammals. Spotila and his colleagues further noted that the metabolic rates recorded for leatherbacks involved in different activities at various temperatures were intermediate between those expected from a typical reptile and a typical mammal.

Do turtles have teeth?

No turtles alive today have teeth; however, teeth have been found in some of the earliest fossil turtles. The teeth have mostly been small ones in the roof of the mouth, and not the standard set of biting and chewing teeth of the upper and lower jaws characteristic of most other vertebrates. Although true teeth are absent, some turtles have serrated areas on their jaws that function like specialized teeth. For example, the front of the upper jaw of some cooters (*Pseudemys*) is jagged, making it easier for them to tear underwater vegetation during feeding. Despite the absence of teeth, the turtles' mouths vary according to their feeding habits. The upper and lower jaws of diamondback terrapins (*Malaclemys terrapin*) and many of the map turtles (*Graptemys*) are broad and flat for crushing mollusks (snails and clams) on which they feed. Both upper and lower jaws of alligator snapping turtles (*M. temminckii*) are beaklike for puncturing or tearing large fish or other animals. Most turtles have relatively sharp-edged upper and lower jaws suitable for slicing plant material, animal prey, or occasionally the finger of a careless biologist.

Do turtles sleep?

Turtles are presumed to sleep in the same manner as other vertebrates. Most turtles are active during the day or at dusk but are inactive at night, at which time they retreat to a refuge. Turtles close their eyes as they rest or sleep and sometimes as they bask during the day. Many of the freshwater turtles captured during research at the Savannah River Ecology Laboratory and kept overnight in the laboratory in shallow water in bins actually extend all four legs while sleeping, but whether this is a common behavior in the wild has not been reported. However, several species extend their legs while basking, and some presumably doze off while doing so.

Can turtles see color?

In addition to having rods (which are the light receptors) in the eyes for basic monochromatic vision, as would be expected, those turtles that have been examined also have cones, indicating that color vision is possible. The extent and pattern of color perception among turtle species, or even within any single species, is still not specifically known. Nonetheless, at least some turtles are assumed to be able to see some parts of the color spectrum. Adult males of some species (for example, the painted terrapin [*Callagur borneoensis*] of the Malaysian region) not only have coloration that differs from adult females but also may change color during the breeding season. Another example is the red eye color of adult male box turtles (*Terrapene carolina*), which is thought to be a biological cue among breeding adults that allows an individual to distinguish between the sexes. Although turtles may not see the same color spectrum as humans, at least some are able to detect colors outside of the grayscale. Carl Ernst, now a retired professor of biology from George Mason University, conducted experiments to determine whether turtles preferred certain colors over others. He concluded categorically that "color vision exists in turtles." His findings showed that species from four different North American turtle families differed in their choices of colors. Snapping turtles (*Chelydra serpentina*) were more likely to choose blue light, whereas spiny softshell turtles (*Apalone spinifera*) preferred yellow light.

Nonetheless, despite Ernst's interpretations, the choice of particular colors does not necessarily mean that any of the turtles tested were able to differentiate among a broad spectrum of color. Because of differing ecological niches (or lifestyles) in which color may be important in some species and less so in others, some turtle species are probably able to detect a wider range of color than others. Some yet to be studied are likely to have monochromatic vision. For example, snapping turtles have no coloration on the shell or body parts that would aid the sexes in differentiating between each other during courtship, and they do not have a diet that specializes in brightly colored plants or animals for which color detection would be an advantage. They may prefer blue light to yellow, red, or green when given a choice because blue appears darker, not because it is the color blue. Further scientific experiments will need to be conducted before a definitive statement can be made about color vision in most species of turtles.

Tortoises have been reported to be strongly attracted to red, orange, and yellow flowers and fruit, which they can spot at distances of several yards while foraging. One of the most dramatic displays of color recognition in our experience was that of an adult gopher tortoise during a presentation to a middle school class. Upon being placed on the floor of the gymnasium during the talk, the tortoise was fed a bright red strawberry so the children could watch it eat. After eating the fruit, the tortoise looked across the floor and began lumbering toward a sixth grade girl who suddenly realized that she was the target. She had toenails painted bright red, and the tortoise clearly had focused on them as his next snack.

Do all turtles have hard shells?

Not all turtles have hard shells, and among those that do, the thickness of the shell varies considerably. The family of softshell turtles has leathery shells covering the bony carapace and plastron. Sea turtles have hard, thick protective shells, except for the leatherback sea turtles (*D. coriacea*), which are covered with a thick, oily material with a leathery surface. The contrast in shell thickness often indicates the array of predators a species confronts. For example, the shells of slider turtles (*Trachemys*) and Florida red-bellied cooters (*Pseudemys nelsoni*), living in southeastern areas of the United States where large alligators are abundant, are very thick. In contrast, chicken turtles (*Deirochelys reticularia*), which generally inhabit small seasonal wetlands where alligators are infrequent inhabitants, have relatively thinner shells. Terrestrial species, such as box turtles (*Terrapene*) and tortoises that

Turtles in the family Trionychidae have pliable, leathery shells rather than the hard-shelled protective carapaces of most other turtles.

Photo © Mike Plummer

are likely to encounter mammalian predators on land, have tough, thick shells that are difficult to penetrate. The effectiveness of the thick armor of slider turtles (*Trachemys scripta*) on barrier islands of the U.S. Atlantic coast, where large alligators are present, has been noted, and live specimens are frequently found with alligator tooth marks and occasionally with alligator teeth embedded in the carapace or plastron.

Can a turtle emerge from its shell?

Despite the ability of turtle cartoon characters to zip in and out of their shells when they are in a hurry, real turtles can do no such thing. A turtle leaving its shell behind would be like a dog strolling down the street while leaving its ribs and backbone on the porch. Hermit crabs, a common crustacean of coastal areas, can actually leave their shells, but in reality the shell is not their own. It formerly belonged to a mollusk. Hermit crabs are not physically attached to the shell but simply use it as protective housing (sort of the mollusk version of a mobile home, which is discarded and replaced as the crab grows). With turtles, the shell is an integral part of the body that is attached with muscles, cartilage, and bone, and it grows as the turtle grows.

Can turtles run?

Most turtles walk at a slow, steady pace when on land, but some can move much more quickly than one might expect. When a gopher tortoise (*Gopherus polyphemus*) is threatened by an approaching person, it will move through an open pine woods habitat at a speed of several miles an hour to reach its burrow. Cooters (*Pseudemys*) and slider turtles (*T. scripta*) also sprint rapidly toward the water if they are startled while on land, as will an

Turtles: The Animal Answer Guide

With turtles the shell is an integral part of the body that is attached with muscles, cartilage, and bone and grows as the turtle grows. Finding an empty shell is a sure sign that the turtle has died. If the turtle has been individually marked, as with this shell, the animal's individual identification can be determined after it has died, providing information on distance moved or increase in body size since last capture. Photo © Whit Gibbons

arrau (*Podocnemis expansa*) on a river nesting beach. However, no turtle can outrun even a slowly jogging human.

Can all turtles swim?

Some turtles, namely, tortoises, cannot swim, although some may float passively for hours or possibly days and reach shorelines many miles from their point of origin. Nonetheless, the terrestrial tortoises characteristically avoid bodies of water, and adult gopher tortoises (*G. polyphemus*) have been known to sink to the bottom and drown after falling into man-made lakes. Most North American box turtles (*Terrapene*) are also primarily terrestrial but will sometimes enter wet areas during drought (they seem to enjoy a good soak in the mud occasionally) and can swim, albeit slowly, across deeper waters in a lake. Coahuila box turtles (*Terrapene coahuila*) as well as Gulf Coast box turtles (a subspecies of the common box turtle, *T. carolina major*) have been reported to mate in the water.

Do all aquatic turtles have flippers?

All sea turtles have flippers, as do the freshwater pig-nosed turtles (*Carettochelys insculpta*) of northern Australia and New Guinea. However, other aquatic turtles have webbed feet that make swimming easier (to which any human who has worn artificial flippers can attest). In some highly aquatic groups, such as many of the softshells (*Apalone, Rafetus, Trionyx*), all 4 feet are fully webbed. In species such as the mud turtles (*Kinosternon*) and sliders (*Trachemys*) that are semiaquatic and spend parts of their life cycle on land, the webbing, especially on the front feet, is reduced.

Form and Function

Movements of an adult painted turtle with a small radio-transmitter attached to the shell are not impeded by the device. Radio-telemetry has been an aid to turtle biologists in determining where turtles spend their time. Photo © Justin Congdon

Can turtles breathe underwater?

All turtles have two well-developed lungs and typically breathe through their noses. A turtle trapped underwater will drown within hours if it cannot reach the surface to breathe. However, the time it will be able to survive without oxygen depends on the turtle's body temperature and rate of metabolism. At cold water temperatures, a turtle's metabolism is lowered, allowing it to survive longer underwater.

Some species of turtles can obtain limited amounts of oxygen while underwater through vascular tissue located in the mouth or the cloaca. This mechanism is most effective during cold periods when the animal's metabolism is so low that very little oxygen is required. One of the most impressive mechanisms for underwater respiration among turtles is that of the Fitzroy River turtle (*Rheodytes leukops*). The species is the only one in its genus and was discovered in Australia in the 1970s and described in 1980 by John Legler of the University of Utah and John Cann, author of *Australian Freshwater Turtles*. In essence, the Fitzroy River turtle was able to breathe underwater with gill-like structures located within the cloaca. The species is known only from the single river system, which is replete with riffle areas that are highly oxygenated by flowing waters.

A study conducted by Tony Tucker of Queensland Parks and Wildlife Service, Australia, and several colleagues in consideration of the underwater respiration by the species revealed the importance of understanding both ecology and physiology of a turtle species in addressing conservation concerns. By tracking radio-telemetered Fitzroy River turtles, the scientists were able to determine that individual turtles traveled around in the river

The Florida softshell turtle has a distinct proboscis that allows a turtle sitting in shallow water to breathe at the surface without being noticed by predators or prey. Photo © Kurt Buhlmann

system but always stayed near highly oxygenated riffle areas. The cloacal respiration discovered in this species was presumably a critical feature of its ecology and behavior, either of which could be detrimentally affected by the construction of dams that reduced oxygen levels in the river water.

Can you tell whether a fossil turtle lived in the sea, in fresh water, or on land?

The shapes of the shell and bones of the feet, which are often preserved in fossil material, are excellent clues for determining whether a turtle's general habitat was land or water. A turtle with a high-domed shell and thick, columnar feet was almost certainly terrestrial. A turtle with a relatively streamlined shell and foot bones that splayed out into broad, flat flippers was certainly aquatic and probably lived in the ocean. Geologists are able to determine whether the specific habitat was fresh or salt water by the chemical composition and other characteristics of the sediments. Turtles today with intermediate shells between highly domed and streamlined are usually semiaquatic, often spending as much time on land as in the water. The same traits would have applied to ancient turtles, hence shell shape can even indicate how a turtle's life was divided between a terrestrial and an aquatic existence. When paleontologists remove fossil remains from a deposit whose geologic age and ancient habitat (for example, a coastal sea, freshwater marsh, or terrestrial upland) have already been determined, an additional clue has been provided for whether the turtle lived in the sea, in fresh water, or on land.

Chapter 3

Turtle Colors

Why do so many turtles have yellow stripes on their neck?

Yellow stripes on the head, neck, and legs are the single-most common color traits among turtles of the world, and many species of North American and Asian turtles are noted for their distinctive yellow-striped patterns. The simplest explanation for yellow striping is that it is camouflage for species that live in vegetated wetland habitats where the stripes can be easily mistaken by predators for strands of aquatic weeds. The tendency for young turtles of many of the species with yellow neck stripes to stay in shallow areas of floating vegetation in which stems are often yellow attests to the safe haven provided by these similarly yellow-mottled habitats. Many species do not have yellow stripes but instead have yellow spots (distinct or indistinct) on the carapace, head, or limbs.

The unique patterns of stripes and other yellow markings often aid not only turtle biologists but also the turtles in species recognition. For example, the eastern mud turtle (*Kinosternon subrubrum*) has a solid black head in areas where its geographic range overlaps with the striped mud turtle (*Kinosternon baurii*), which has two very distinct yellow stripes on each side of its head; therefore, when species recognition is important, perhaps when a turtle is looking for a mate (admittedly, not a good time for an error!), the two can readily distinguish each other by appearance. Interestingly, the common mud turtle has yellow stripes in portions of its range in the western parts of Alabama, Mississippi, and Louisiana, in areas where the striped mud turtle does not occur. In some species, especially slider turtles (*Trachemys scripta*), the yellow stripes fade with age, and in adult males, stripes on

Many freshwater turtles have yellow stripes on their head and neck. The striping pattern is distinctive for some species, such as the common map turtle (*Graptemys geographica*; *left*) and Florida red-bellied cooter (*Pseudemys nelsoni*; *right*). Photos © Kurt Buhlmann

the head and legs can disappear completely, becoming dark greenish gray or black, a condition known as melanism.

What causes the different shell and skin colors of turtles?

Overall color patterns of turtles' shell and appendages help individuals blend into their natural habitats, an advantage to individuals both for capturing prey and avoiding predation. Although a box turtle (*Terrapene carolina*) crossing a road may appear colorful against the background of asphalt, the dappled carapace of the same individual would blend in well in its natural habitat—a wooded area with streaks of sunlight on a ground surface covered with yellow and brown leaves. The dark shells of mud turtles (*Kinosternon*) and snapping turtles (*Chelydra*) match the muddy substrate of lakes and pond bottoms where they walk in search of prey. The pancake-colored leathery carapaces of softshell turtles (*Apalone*) are ideal camouflage on the bottom of sandy rivers where many dwell. Color may vary among individual species, but those that blend in with their environment are most likely to survive longer and so be favored through natural selection.

A turtle with a camouflaged shell can avoid detection by potential prey as well as predators and may actually go unseen by prey animals. A dramatic example is the matamata (*Chelus fimbriata*) of South America whose dark skin and carapace allow it to sit motionless in the water until small fish approach, whereupon it opens its mouth and sucks in the unsuspecting prey. Its color pattern of black and brown matches the dark and muddy waters where it lives. Another use of color for prey capture is the bright pink

Any species of animal can have an albino in which all black pigment is absent. The solid white mud turtle with pink eyes is one of two captured by researchers at the Savannah River Ecology Laboratory during studies in which more than a thousand mud turtles with normal pigment were captured. SREL file photo

tongue of the alligator snapping turtle (*Macrochelys temminckii*). It is used as a lure against the contrasting brown and black colors of the turtle's open mouth to coax unsuspecting fish into range to be quickly "snapped up," or eaten, as the turtle lies motionless on the bottom of a river or lake.

The physiological causes of light and dark coloration that can occur within individual turtles of some species are a consequence of the level of concentration of melanophores that produce dark coloration and chromatophores (pigment cells) that produce reds and yellows. A class of chromatophores known as iridophores are present in some reptile species, particularly lizards and snakes but also in some turtles. These iridophores appear white or reflect light to create a blue appearance.

What color are a turtle's eyes?

Eye color is a defining characteristic among and within some species. Most turtles have yellow or yellowish brown eyes with black pupils, and the eyes of most of the remaining species are shades of brown or black. Male common box turtles (*T. carolina*) and spotted turtles (*Clemmys guttata*) often have eyes that are reddish in color in contrast to the brown or brownish yellow eyes of most females. The two subspecies of the false map turtle (*Graptemys pseudogeographica*) differ in eye color. The midwestern false map turtle (*G. p. pseudogeographica*) has a yellowish brown iris and a horizontal black bar running through the pupil, whereas the Mississippi map turtle (*G. p. kohnii*) is noted for having a white iris, black pupil, and no bar. Adult males of the mangrove terrapin (*Batagur baska*) of India and Southeast Asia have a light-colored iris that appears white or light yellow in contrast to the

Turtles: The Animal Answer Guide

adult females' brown eyes. The brown roofed turtle (*Pangshura smithii*) of India can actually be viewed as having beautiful blue eyes, and Jeff Lovich, coauthor of *Turtles of the United States and Canada*, noted that a subspecies of the Spanish terrapin (*Mauremys leprosa*) in Morocco also has blue eyes.

The configuration of the pupil and associated pigmentation varies considerably among species, usually appearing as a distinct black circle in most species but as a horizontal bar in many others, including chicken turtles (*Deirochelys reticularia*), slider turtles (*T. scripta*), and the Chinese stripe-necked turtle (*Ocadia sinensis*). The pupil is star-shaped in the alligator snapping turtle (*M. temminckii*). Malaysian giant river turtles (*Orlitia borneensis*) have brown eyes with an outer circle of blue iris.

Is there a reason for the patterns on the shell?

Some turtles have intricate shell patterns on the carapace and plastron with such consistent designs within species that they are used in identification. Others have shells that are virtually monochromatic, usually in basic black or brown. In nearly all cases, the shell colors and patterns protect the turtles in their preferred habitats by acting as a form of camouflage from predators. The hieroglyphic-type swirls on the carapace of a young cooter (*Pseudemys*) may look like ripples on the water surface to a wading bird that would readily eat a small turtle. The yellow, orange, or red rings on the ringed map turtle (*Graptemys oculifera*) may also afford protection for juveniles susceptible to birds of prey. Contrasting patterns on the plastron, such as the black blotches on the yellowish undersides of a red-eared slider (*T. s. elegans*), may make it difficult for aquatic predators such as gar or catfish to detect the turtle as it swims on the water's surface above them.

Do a turtle's colors change as it grows?

Newly hatched turtles are often more brightly colored and sometimes patterned differently than older individuals of the same species and remain so for a year or more, gradually becoming more muted or darker as they age. In a few species such as painted turtles (*Chrysemys picta*) of North America, the babies are miniatures of the adults with little difference in color. Some adult painted turtles are as colorful as when they were hatchlings. However, an adult eastern mud turtle (*K. subrubrum*) with a plain dark brown carapace and dull yellowish brown plastron begins life with a black carapace and bright red or orange plastron with a few black streaks. During the first year, the bright color fades into the typical yellow and brown of the adults. In the striped mud turtle, the underside of hatchlings is yellow or black. Baby gopher tortoises (*Gopherus polyphemus*) have large

A baby gopher tortoise is more brightly colored than the dull gray of the adult. Photo © Trip Lamb

yellow areas on the carapace that fade with age, so that old adults appear dull yellowish or brownish gray.

Although young turtles are often more strikingly colored than adults, exceptions occur. Baby box turtles (*T. carolina*) have a more muted color pattern than the adults, a trait that offers better camouflage for a tiny animal on the forest floor. A box turtle hatchling with a dirt brown shell and pale yellow spots down the center of the carapace and on the scutes has a very different appearance from the bright yellows, blacks, and oranges of the adult carapace.

A distinctive color change that occurs as turtles age and grow larger in some species, most notably the slider turtle (*T. scripta*), is increasing melanism. Adult male slider turtles become melanistic as they get older and larger. They sometimes completely lose all of the yellow coloration on the head, limbs, and shell, including the yellow or red spot on the side of the head characteristic of yellow-bellied (*T. s. scripta*) or red-eared sliders (*T. s. elegans*). The most thorough analysis of the phenomenon of melanism in turtles was in the chapter "Development and Significance of Melanism" by Jeff Lovich (U.S. Geological Survey), Jack McCoy (deceased; Carnegie Museum), and Bill Gartska (University of Alabama at Huntsville) in *Life History and Ecology of the Slider Turtle*. In an examination of thousands of slider turtles, they concluded that melanism is a gradual change from brightly colored immature male sliders to the ultimately dark gray or black patterning associated with older individuals. Variability in males within the study populations was high but did not typically occur until males reached sizes at which females reached maturity, leading researchers to suspect that melanism may aid in mate selection during courtship. Since the darker coloration is also believed to be associated with hormonal changes that occur with the onset of maturity, an interesting study would be to deter-

Dark pigment concentrates in male slider turtles as they get older and larger so that they lose the colorful stripes on the head and legs and the yellowish pattern on the carapace and become melanistic. Photo © Cris Hagen

mine whether the darker males are preferred by females and have more offspring.

Do a turtle's colors change in different seasons?

The color patterns of almost all of the world's turtles remain constant throughout the year, although the colors of a few species change in different seasons. The most dramatic seasonal color changes in a turtle occur in painted terrapins (*Callagur borneoensis*) of Malaysia, Sumatra, and Borneo. In this species, the color of the usually nondescript gray head and dull reddish stripe down the center of the top of the head of the males changes drastically during the breeding season, with the head becoming white and the stripe becoming bright red. The upper lip is black. The carapace of male river terrapins (*B. baska*) also changes color during the breeding season, becoming darker, almost black, rather than the usual gray. Other species of turtles may also change color seasonally, but the differences are less dramatic and more difficult for humans to quantify. Of course, turtles know what they are looking for in a mate and may be more impressed by subtle color differences than their human observers are.

Is there much geographic variation within a single turtle species?

Geographic variation is apparent in several species of turtles that are widely distributed with some differences significant enough that distinct subspecies have been recognized. One of the best-known examples is the slider turtle (*T. scripta*), which has three subspecies in the eastern United

States. The yellow-bellied slider turtle (*T. s. scripta*) in the eastern part of the range has a yellow blotch or broad stripe on each side of the head behind the eye, in contrast to the red-eared slider (*T. s. elegans*) in which the markings behind the eye are bright red. Another difference in appearance is that yellow-bellied sliders typically have a clear yellow plastron with two black spots at the front while red-eared sliders have a yellow plastron with numerous black spots from front to back, usually one per scute. In the Cumberland slider subspecies (*T. s. troostii*), the stripes on the side of the head are yellow but are narrower than in the yellow-bellied slider. All three subspecies of U.S. sliders are similar in ecology and behavior.

Common box turtles (*T. carolina*) are partitioned into four subspecies that vary noticeably in color pattern and to some degree in shell morphology. The true geographic variation of the striped mud turtle (*K. baurii*) was unknown for many years. In the original description of the species, the bright yellow stripes down the center and sides of the carapace were a signature character for the species. Although the striped mud turtle was known in much of Florida and parts of southern Georgia, more than a century passed before the species was identified in the Carolinas and into Virginia. Confusion resulted because the more northern populations have pale or nonexistent stripes on the carapace. The lack of geographic variation in most turtle species occurs because they exist in a fairly limited area or territory. For example, most of the map turtle species are restricted to one or a few adjacent river drainages.

Among the most extensive geographic variation in a single species is that of the diamondback terrapin (*Malaclemys terrapin*) that has a narrow linear distribution from Cape Cod down the East Coast, around the peninsula of Florida all the way to Texas in the estuarine area of coastal salt marshes. Seven subspecies have been described, and regional variation is obvious in some of these. For example, the mangrove terrapin is different in appearance than those in either end of the distribution. However, the variation in color pattern among the subspecies of terrapins occurring along the Atlantic Coast is often as great within a single coastal population as it is among turtles classified as different subspecies because of their color patterns.

Chapter 4

Turtle Behavior

Are turtles social?

With the exception of interactions during the breeding season, most turtles are relatively nonsocial animals during all life stages. Social interaction among males and females, of course, occurs during courtship and mating, and males of some kinds of turtles engage in combat during the mating season. Both sexes of diamondback terrapins (*Malaclemys terrapin*) may congregate in large groups of several dozen or possibly hundreds during the mating season as well. Such congregations may occur for other species, but turtle biologists have not documented the phenomenon. Other situations in which turtles, including cooters (*Pseudemys*), map turtles (*Graptemys*), and sliders (*Trachemys*), are observed together in large numbers are when basking on logs or rocks in rivers or lakes. However, this is not true social behavior because interactions among the animals usually occur independent of other individuals. During basking, turtles will behave toward one another no differently than if they were competing with a wading bird or small alligator for valuable space on the log or rock.

Occasionally, turtles competing for the same basking sites may be overtly aggressive to others of the same species. Bruce Bury with the U.S. Fish and Wildlife Service and Jaclyn Wolfheim at the Smithsonian's U.S. National Museum reported aggressive behavior among Pacific pond turtles (*Actinemys marmorata*) basking on rocks in a natural situation. As a student at George Mason University in northern Virginia, Jeff Lovich studied basking painted turtles (*Chrysemys picta*) and observed more than 200 incidents in which turtles bit, pushed, or gave open-mouthed gestures to other individuals basking on the same log. Other species of turtles may exhibit

Giant tortoises on the islands of the Galápagos and Aldabra often congregate in shaded areas or pools of water. The behavior is not necessarily of a social nature but may simply represent an area with the most suitable environmental conditions. Photo © Judy Greene

the same behavior, but observers with the patience to record such activity have been few. Such behavior may be of minor importance within turtle populations under natural conditions, although in areas where basking sites are limited, such territoriality may be biologically significant.

Gopher tortoises (*Gopherus polyphemus*) appear in some situations to be social animals because they live in well-defined colonies similar to highly social prairie dogs. Indeed, some biologists, including Dr. Tracey Tuberville of the University of Georgia's Savannah River Ecology Laboratory (SREL) who has studied several gopher tortoise colonies, believes that these tortoises exhibit social behavior. They may prefer particular burrows as well as other particular tortoises, visiting their "friends" in other burrows in a nonrandom pattern. Further research, including continued behavioral observations, will be needed to determine the influence of social behavior in the clustering of tortoises in burrows and in the placement of burrows on the landscape. The juxtaposition of burrows is certainly a function of environmental conditions suitable to the individual for digging, feeding, and basking but social behavior may also be a factor. Gopher tortoise colonies

Turtles: The Animal Answer Guide

definitely exist on open, longleaf pine sandhills with extensive ground vegetation because of the suitability of the habitat, but the distribution and proximity of burrows are presumably the consequence of a social function as well.

Other major congregations of turtles occur during the massive egg-laying sessions of ridley sea turtles (*Lepidochelys*) and the arrau river turtle (*Podocnemis expansa*) in the Orinoco River of Brazil. Before the drastic, some would say catastrophic, decline of these once-abundant species, each species arrived synchronously at nesting beaches at which thousands of females of their species would lay eggs at the same time. A mass nesting of ridley sea turtles of either species is called an *arribada*. The first arribada recorded on film was in the early 1960s on a beach in the Gulf of Mexico where an estimated 40,000 female Kemp's ridleys (*Lepidochelys kempii*) nested during a single day.

One biological interpretation of the mass nesting phenomenon is that it is not really social behavior but a tendency of individuals to go ashore independently when environmental conditions are the most propitious, as with turtles basking on logs or tortoises developing colonies in a prescribed area (or even humans heading en masse to the beach on the first warm spring weekend). However, some turtle biologists suggest that nesting in large numbers over a short period is a strategy to thwart egg predators by satiating them, which lowers the probability that an individual female's nest will be predated and ensures that many eggs will hatch successfully.

Do turtles fight?

The primary reason for a turtle to engage in combat with another turtle is to compete for a limited resource. Receptive females are such a resource, and male-male combat among individuals has been documented within several species. Males of common snapping turtles (*Chelydra serpentina*) and some tortoises are known to fight other males of their species during mating season. Presumably, the winner of such fights is rewarded by being the one to successfully mate with nearby females. Although male turtles may fight in such situations, significant injury is seldom apparent after combat, although a tortoise turned upside down with no way to right itself, which may rarely happen, could result in death.

However, as with many ritualistic behaviors among animals in which males challenge one another, the loser leaves the area but is usually not permanently harmed. Interestingly, some observations originally reported as snapping turtles fighting in the water, upon closer observation (by an experienced biologist or a film of the incident), turned out to be the ritualistic mating behavior of a male and a female, with the pair eventually mating.

Turtle biologists must be certain that what appears to be fighting between two males is not actually part of the courtship ritual between the sexes.

A few species of turtles, including snappers (*C. serpentina*), will attack other snappers or other species of turtles when they are confined in proximity with other turtles, such as in a trap, a tank, or an aquarium. This aggressive behavior is also extended indiscriminately to other animals in the trap, such as frogs or fish, and to the trap-checking human, who created the problem to begin with. Whether this is true "fighting" or merely snapping turtles revealing their true grumpy natures is hard to discern.

An example of turtles seeming to fight or at least showing aggressive behavior toward one another is in artificial congregations of turtles being fed. This happens where turtles are held captive in ponds or other nonnatural situations where they are fed regularly together. More jockeying for position than true fighting occurs, but actions interpreted as aggressive have been recorded at these feeding frenzies.

Are snapping turtles the only ones that bite?

Most turtles are capable of biting, as this is the manner in which most of them obtain food, and many or perhaps most species defend themselves by biting. For example, an aquatic turtle removed from the water is clearly in a threatening situation, and biting is a form of self-protection. However, some species of turtles do not bite people or do so rarely. Tortoises are usually inoffensive toward humans and do not usually bite as a form of defense. Box turtles (*Terrapene*), likewise, seldom bite, although an occasional individual will. One amusing incident to all but one person present was in Michigan when a participant named Sue Novak, who was assisting with a turtle research project, encountered a large Blanding's turtle (*Emydoidea blandingii*). When she asked whether they ever bit, we told her that we had handled hundreds of Blanding's turtles and that they were extremely docile and never bit humans. She should have been told that they *rarely* bite people because she met the exception that proves the rule. Being a good scientist she tested the information by putting her finger right in front of its mouth. It grabbed her finger, giving her a decidedly painful pinch, which was no worse than, say, a small snapping turtle (*C. serpentina*) or big slider turtle (*Trachemys scripta*), but as far as we know, it is the only record of a Blanding's turtle ever biting anyone.

As would be expected, the most severe bites to people are from large turtles, including common snapping turtles, alligator snapping turtles (*Macrochelys temminckii*), softshell turtles (*Apalone*), and occasionally sea turtles. All have sharp, cleaving jaw margins, and can deliver a slashing bite that will bruise and draw blood. Alligator snappers are especially formidable

Blanding's turtles are one of the most docile large turtles in the United States and Canada. They are often observed in lakes and marshes because of the bright yellow chin that protrudes above the water's surface. The function of the yellow chin has not been determined. Photo © Kurt Buhlmann

because of their sharp beak, and the largest ones could potentially bite off a person's finger. Large sea turtles will also occasionally bite and, because of their massive size, could cause serious injury. Sea turtle biologists vary in their experiences with these animals, and few have been bitten. Understandably, anyone is likely to be cautious around such enormous animals and not offer an opportunity. Some consider loggerheads (*Caretta caretta*) to be the most likely of the sea turtles to bite. If you put a hand purposefully or carelessly in front of a nesting female, she may bite. Other sea turtles do not have reputations for intentionally biting people. Nonetheless, one turtle biologist told the story of the severed head of a ridley sea turtle that was still alive and actually bit someone's finger off, but we assume that this was the result of involuntary muscle reflexes. Among the species of turtles that seldom bite are spotted turtles (*Clemmys guttata*), wood turtles (*Glyptemys insculpta*), most tortoises, and Blanding's turtles (with one exception). A matamata (*Chelus fimbriata*), the South American turtle that feeds by vacuuming in quantities of water and sediment to capture fish and other aquatic animals, would probably not be capable of delivering an effective bite to a person.

How smart are turtles?

Measuring the intelligence of turtles is a difficult task because, as humans, we tend to associate learning and the capability of learning with traits and characteristics that may be of no innate value to other animals. Thus, many turtles live in aquatic habitats and need a skill set entirely foreign to most humans to survive. That they do not respond like primates or other mammals might make some people consider them dumb. Turtles can, however, definitely learn from previous experience, especially when it involves a behavior that could influence survival. For example, turtles living in a pond can quickly learn to approach a feeding station once they associate a particular action, such as someone splashing the water, with the presence of food. P. K. Greene, a resident of Jackson, South Carolina, found that many slider turtles (*T. scripta*), as well as his pet catfish, would somehow sense the arrival of his pickup truck at his farm pond (probably by sensing ground vibrations) where he fed them almost daily in the warm months. Someone stationed on the shore could see the boiling water caused by the fish and turtles before he even parked and got out of the vehicle.

One of the most dramatic examples of turtle learning behavior involved a common box turtle (*Terrapene carolina*) that lived behind the living room couch in the apartment of a graduate student at Michigan State University. When friends visited, the turtle remained in the same spot until its owner, Sig Nelson, tapped on a glass bowl with ice cream in it. The turtle would be heard rumbling behind the couch as it came out to the center of the living room to approach the person holding the bowl. Standing on the floor beneath the bowl, the turtle would push upward with its front feet until it balanced itself on the back of its shell and its hind feet. The box turtle literally stood up in this manner until it was given a bite of ice cream with a spoon. The bowl was then put down for the turtle to eat the rest of the ice cream.

Sea turtles are also able to find their way back to the same nesting beaches or ocean feeding areas after traveling for many years and covering thousands of miles in the open ocean, a feat that most humans would have a hard time accomplishing without modern technology and equipment. Freshwater turtles of many species also return to the same area to nest every year, some traveling for several miles. Some individuals, most often males, move back and forth between bodies of water that may be several miles apart. During drought periods, many species of turtles from isolated wetlands head directly toward the nearest permanent body of water, although the new aquatic habitat may be hundreds of yards away and is usually not visible from the drying site. Studies have shown that they "know where they are headed" and do not just walk helter-skelter in any direction as the water levels decrease.

Do turtles play?

The most obvious behaviors that can be construed as play in nature are normally associated with social animals, such as dolphins, wolves, and primates. Wild turtles have not been reported to exhibit behaviors that could be described as play. However, Gordon Burghardt of the University of Tennessee and colleagues observed and videotaped what they concluded was play behavior in a Nile softshell turtle (*Trionyx triunguis*) at the National Zoo in Washington, DC. The turtle would approach a basketball that was placed in its aquarium tank and bump it with its nose and bite at it. When the turtle was fed (live goldfish dumped into the tank), it completely ignored the two basketballs that floated on the surface. When food was not present, the turtle interacted with one or the other of the basketballs for more than 3% of the time that a basketball was available. Play behavior by the turtle was observed even more frequently with a floating rubber hoop that occupied the turtle's attention more than 12% of its time. The turtle would not only nose, bite, and shake the hoop, but it would sometimes swim through it, turn around in the water, and swim back through it.

We observed an odd phenomenon with a young yellow-bellied slider turtle (*T. s. scripta*) kept in an outdoor tank at SREL that we concluded was play behavior. A board in the middle of the tank had been placed as a slanting basking platform and was elevated about a foot above the water's surface. While we were testing a remote video camera, we observed the behavior of several turtles in the tank when no one was present. A particular turtle was taped one night as it swam to the end of the board in the water, walked up to the end of the plank, and then dove off of the extended end into the water. It immediately swam around to the lower end, climbed up again, and repeated the process. The little turtle did this nonstop for at least 10 minutes. We found no other explanation than that the individual was "playing" but have no biological reason why it might have done this as it did not involve food, mates, or escape from a predator.

Cris Hagen of SREL had a captive-raised female northern diamondback terrapin (*M. terrapin*) that was around 10 years old. For whatever reason, he put a plastic finger puppet shaped like a dolphin in her aquarium. For 2–3 months, the terrapin showed great interest in the finger puppet and spent a significant amount of activity time "playing" with the puppet. She would push it around the tank with her snout, wear it over her head and swim around, and push her head in and out of it. This behavior continued for several months.

Do turtles talk?

A few turtles vocalize and some communicate by other means, but no turtles communicate in a way that humans would define as talking. Some tortoises make grunting sounds usually associated with aggressive behavior toward other individuals or made between a male and female during mating. The big-headed turtle (*Platysternon megacephalum*) of Asia will some-

Some tortoises make grunting sounds associated with communication between a male and female during mating.

Photo © Cris Hagen

times emit a loud and impressive growling sound when removed from the water. The sound, akin to a roar, is almost prehistoric and would dissuade some predators from trying to make a meal of the turtle. Ironically, a threatening sound seems almost unnecessary because if a large big-headed turtle, with its enormous mouth, were to bite an attacker, the consequences would be severe. Another vocalization, reported by Dick Vogt, in arrau turtles of South America (*P. expansa*), is that they emit underwater sounds that he concluded were a form of long-distance echolocation. The behavior would be especially effective in the wide, dark waters of the Amazon River where the turtles move long distances while underwater.

Turtles also have a variety of forms of silent communication. Males of some aquatic species such as painted turtles (*C. picta*) and slider turtles (*T. scripta*) use their long foreclaws to attract mates during courtship in a behavior known as titillation. Some of the tropical American sliders do not have long claws but instead have long noses through which they force a stream of water into the face of the female while courting. When he was a graduate student at SREL, Jeff Lovich observed rhythmic blinking of the eyes underwater by female slider turtles facing male turtles. The female's eyelids were a lighter color than the surrounding area of the head, and Jeff interpreted the telegraphic blinking as a form of communication behavior related to courtship since it was only observed in the presence of pursuing males.

Male gopher tortoises (*G. polyphemus*) will communicate at the mouth of an underground burrow when they know another animal is in the burrow

Turtles: The Animal Answer Guide

by pounding the plastron against the ground repeatedly. The act is a precursor to mating and is presumably intended to bring the female tortoise in the burrow to the surface. According to Tracey Tuberville of SREL, who has studied the species extensively, the behavior may also be used by either sex to determine whether a burrow is unoccupied and therefore available for use. When another tortoise is already in the burrow, the occupant will often come to the entrance and hiss at the other tortoise or block the burrow entrance. A person pounding on the ground in front of the burrow can sometimes "call" a tortoise to the surface.

The common expression "voice of the turtle" in the Song of Solomon 2:12 from the King James version of the Bible does not refer to a turtle but instead to a bird, the turtle dove.

How do turtles avoid predators?

The first step for any prey species is to avoid the predator entirely or to remain undetected when in proximity to the predator. Among the more obvious avoidance behaviors are those found in gopher tortoises (*G. polyphemus*), which retreat to their burrows, and African pancake tortoises (*Malacochersus tornieri*), which lodge themselves into rock crevices out of the attacker's reach. Flattened musk turtles (*Sternotherus depressus*) in Alabama will also retreat to rock crevices underwater where they are less likely to be seen and extracted by aquatic predators. Many aquatic species accomplish the same purpose by swimming beneath the bank or under logs or clumps of vegetation to stay hidden.

A strategy used by some turtle species that live in seasonal wetland habitats that occasionally dry up is to emerge from the basin before all standing water is gone. An individual will dig a small burrow or crawl beneath leaf litter from a few to several hundred feet away from the margin of the aquatic area. Behavior of this sort, which is practiced by many turtles in the southeastern United States, avoids the threat of predators such as otters and raccoons that enter the receding aquatic habitat in search of prey animals that have been concentrated because of the drop in water level. Once they are detected by a predator, some turtles bite and use their claws in defense. The North American box turtles (*Terrapene*) use the highly effective approach of closing up completely in their shell (like a tightly closed box) so that neither their head nor any of their appendages are visible to the outside. (Occasionally box turtles kept in captivity and overfed become too chubby to close their shells.) Many of the large species of turtles reach sizes that ensure they will be predator proof from most natural predators, except some of the large crocodilians where they coexist. Humans are the main threat to the survival of adult turtles.

Some turtles avoid the initial encounter with predators by simply having excellent camouflage that allows them to stay undetected. An example is the spiny softshell turtle (*Apalone spinifera*). These leathery shelled, flat-bodied creatures are the color of sand and live in clear, sandy creeks. They shuffle themselves completely beneath the sandy surface so that they are practically invisible. Soft mud at the bottom of lakes and rivers is similarly used by darker-colored species when they want to hide.

A final resort of many turtle species, such as the common musk turtle (*Sternotherus odoratus*), after being captured by a person or predator is to release a foul-smelling musk from glands located on the underside of the carapace. Emitting the musk is presumably a behavioral attempt to discourage any creature that might want to eat a small turtle (or even a large one, as common snapping turtles [*Chelydra serpentina*] also emit a strong, unpleasant smell), but its effectiveness as a retardant to predators by these and several other turtle species has not been thoroughly investigated.

Chapter 5

Turtle Ecology

Where do turtles sleep?

Turtles sleep in places likely to be safe from predators and from extreme environmental temperatures. For a gopher tortoise (*Gopherus polyphemus*), this means sleeping in the burrow where temperatures remain relatively constant and where the turtle is inaccessible to most predators. The African pancake tortoises (*Malacochersus tornieri*) position themselves safely in rock crevices to sleep. Most freshwater turtles sleep on the bottom or up under the bank of the river, lake, or pond, situating themselves so that they are able to rise to the surface to take a breath of air when needed. A softshell turtle (*Apalone*) hides itself in the sand along a riverbank or along a lake shoreline in water that is shallow enough for it to extend its long neck and snorkel-like proboscis above the surface to breathe as it sleeps. It uses the same behavior when it is awake and waiting for its prey to swim into range.

Many map turtles (*Graptemys*), cooters (*Pseudemys*), and sliders (*Trachemys scripta*) sleep at night on brush piles or tree limbs that place them out of reach of large aquatic predators and far enough from shore to avoid terrestrial predators that prowl the banks. The turtles position themselves within the branches so that they are supported and immobile at a depth that allows them to extend their necks safely out of the water to breathe. The tree limbs do double duty as refuge and alarm because they will vibrate if touched by an approaching predator, serving as a warning system. Basking turtles, which may also be sleeping, exhibit the same behavior. Common box turtles (*Terrapene carolina*) will often conceal themselves beneath a cover of dead leaves or other vegetation to sleep. Although turtles that

In some areas, green sea turtles sleep on the beach during day or night, a behavior that has been observed in Hawaii and Australia.

Photo © Cris Hagen

bask on logs or rocks occasionally doze off, most sleep at night during their inactive period. Sea turtles sometimes sleep in shallow waters on the bottom or while floating immobile in the ocean. In some areas, sea turtles sleep on the beach during day or night, a behavior that has been observed in Hawaii and Australia. The giant tortoises of Aldabra and the Galápagos Islands, where they have no natural predators, sleep in the open with the neck stretched out on the ground.

Do turtles migrate?

Many species of turtles routinely migrate, some over land, some aquatically, for distances from a few yards to thousands of miles (table 5.1). Females of all sea turtles and freshwater species that lay their eggs on land technically migrate from the water to the terrestrial habitat and back during each egg-laying bout. For most species, these are annual events for at least some members of the population and may occur multiple times during the nesting season in some species. Migratory behavior can also be prompted by environmental conditions in which turtles move from a habitat that has become unsuitable, such as drying of a wetland, to a more favorable habitat. They then return to the former habitat when conditions improve.

One of the longest reported migrations by a turtle was a 12,774-mile trek by a leatherback sea turtle (*Dermochelys coriacea*) satellite-tracked by Peter Dutton and Scott Benson of the National Oceanic and Atmospheric Administration's National Marine Fisheries Service's Southwest Fisheries Science Center and Creusa Hitipeuw of WWF-Indonesia. Considering the annual distances traveled by adult leatherbacks, it is not unreasonable

Table 5.1 General factors potentially influencing movements of turtles in a population

A. Environmental	B. Demographic
Daily temperature patterns	Population density
Seasonal temperature patterns	Sex ratio
Weather events	Age structure
Habitat type and condition	Size structure

C. Maturity and physiological state
    Sex
    Body size
    Recent experience

Note: Turtles move from one place to another for a variety of reasons, primarily related to environmental conditions (A); the age, size, and sex within the population (B); and the specific characteristics of individuals (C). A complete summary of the phenomena involved in turtle movement patterns is provided in chapter 16 of the book *Life History and Ecology of the Slider Turtle* (Smithsonian Institution Press, 1990).

to assume that a lifetime accumulation of distance traveled could total in the hundreds of thousands of miles.

How many turtle species live in rivers?

Approximately half of the kinds of turtles in the world can be found in or around rivers and many of the species are found almost exclusively in river habitats. Many of the turtle species inhabiting large rivers are practically never found on land except when females leave the water to lay eggs or when the hatchlings make their way to the aquatic habitat after hatching. In the United States, approximately 20 species are considered to be riverine, including the alligator snapper (*Macrochelys temminckii*), spiny softshells (*Apalone spinifera*), and smooth softshells (*Apalone mutica*), map turtles (*Graptemys*), cooters (*Pseudemys*), and some of the musk turtles (*Sternotherus*).

Although all are found primarily in rivers, most are known to persist in natural or man-made impoundments, and may also occur in river swamps and backwaters. About two dozen U.S. species are classified as semiaquatic, which means they spend most of their lives in seasonal wetlands, small lakes and ponds, and marshy habitats. The majority of these are sometimes peripherally associated with large rivers in slow-moving backwaters, oxbows, tributary streams, or river swamps. Several species of side-necked turtles of Australia, South America, Africa, and Asia, as well as the large softshell turtles of Asia and Africa are river dwellers, and like the U.S. species, these riverine species are often found in large lakes or river backwater habitats.

The pig-nosed turtles of Australia and New Guinea and the Central American river turtle (*Dermatemys mawii*) are all river turtles.

How many turtle species live in lakes?

More than half of the world's species of turtles are capable of living in lakes or ponds, although none is restricted exclusively to natural lake habitats. The freshwater aquatic turtles that live in lakes also inhabit slow-moving areas of rivers, oxbow lakes, man-made reservoirs, drainage ditches, and swamps. Features of any bodies of water that lake-dwelling turtles are likely to inhabit, and which determine the species that are likely to be present, are the permanence of the water, amount of shoreline vegetation, predators that are present, and overall climatic conditions, especially during the winter. Most turtles found in lakes are most likely to be in areas that are heavily vegetated, allowing them to escape certain predators and to find food for themselves, whether plant material or invertebrates, amphibians, or other small reptiles. A similar-sized lake in the same region without vegetation will have far fewer turtles than a lake with floating and submerged plants.

Other features of topography could also influence the suitability of a lake habitat for turtles, such as whether a lake is spring- or stream-fed from intermittent or permanent sources. Accessible streams may provide safe, protected routes for movement to good nesting areas, which are rare around some lakes. Having to traverse long overland distances to nest may prevent some species from thriving in lake habitats.

How many turtle species live in the ocean?

Seven species of turtles are strictly marine. Six belong to the family Cheloniidae: loggerhead (*Caretta caretta*), hawksbill (*Eretmochelys imbricata*), Kemp's ridley (*Lepidochelys kempii*), olive ridley (*Lepidochelys olivacea*), green (*Chelonia mydas*), and flatback sea turtles (*Natator depressus*). The seventh species, the leatherback sea turtle (*D. coriacea*), is placed in a separate family (Dermochelyidae).

Which geographic regions have the most species of turtles?

The areas of greatest species diversity among turtles are the southeastern United States and Southeast Asia. In the United States, more different kinds of turtles (18 species) are found in a few hundred square mile area of the Mobile basin than in comparably sized areas anywhere else in the

Beaver ponds can create aquatic habitat for a variety of freshwater pond species of turtles, such as sliders or painted turtles, that would be less abundant in a stream habitat. Photo © Brian Metts

world. More turtle species are native to Southeast Asia but over a larger geographic area. These two regions indisputably have the highest natural diversity of turtle species on the planet.

How do turtles survive in the desert?

Several species of turtles live in desert regions and have different mechanisms to survive under periods of extreme drought and high temperatures. The most well-known desert turtles are tortoises that have evolved to thrive under environmental conditions that would be lethal for many types of animals, including most other kinds of turtles. An example is the desert tortoise (*Gopherus agassizii*) of the southwestern United States that thrives in the extreme conditions of the Mojave Desert where daily summer temperatures commonly reach over 100 degrees F, lakes and rivers are absent, and rainfall is extremely rare. Desert tortoises remain active from the first warm days in spring until cold weather of late fall and winter.

Tortoises adjust to extreme high temperatures by gradually shifting their periods of activity to the early morning and late evening hours as temperatures grow warmer during the late spring and summer. During the most intense heat, they remain in small burrows that they dig or beneath vegetation or rock ledges that shield them from the sun. Desert tortoises

Desert tortoises survive desert conditions of extreme high and low temperatures as well as weeks or months without rainfall through behavioral and physiological adaptations. Photo © Cris Hagen

are also adapted for hot, dry climates physiologically, by not requiring water beyond what they acquire from vegetation in their diet. They far outdo the legendary camel in that regard. Water retention is greatly enhanced by physiological mechanisms that reduce natural water loss from the body by minimizing evaporative water loss through the skin or from excretion.

Some species of mud turtles (*Kinosternon*) also live under intense desert conditions. For example, populations of the Sonoran mud turtle (*Kinosternon sonoriense*) live in arid areas of Arizona in stream habitats that may experience flash floods and then go for months with no water, eventually drying to intermittent pools that are replenished by rainfall at unpredictable intervals. When confronted with extreme drought situations in which even these pools may dry, the mud turtles survive by moving into the surrounding terrestrial areas where they burrow underground or beneath rocks sometimes for months. Presumably, these turtles have adaptations that allow them to retain water for prolonged periods of time under dry conditions.

Some desert regions have permanent rivers passing through them, such as the Colorado River in the western United States, in which turtles can live as they would in the rivers of more humid areas with moderate temperatures. Interestingly, in contrast to many snakes and mammals, turtles that live in deserts have not developed the behavior of being active and feeding at night to avoid high daytime temperatures.

Turtles: The Animal Answer Guide

Baby yellow-bellied slider turtles hatch in the fall but remain underground in the nest through the winter and emerge at the same time in the spring. They reach the water for the first time as much as a year after eggs are laid. Photo © Whit Gibbons

How do turtles survive the winter?

Most of the world's turtles are found in the temperate zones of North America and Asia and are able to survive typical winter temperatures that may remain below freezing for several days or weeks at a time. Some species are even able to survive the prolonged periods of extreme cold of Canada, northern Europe, and northern Asia. For turtle species to persist under natural conditions in areas with harsh winters, all life stages from egg and hatchling to older adults must be able to withstand the extreme temperatures. For the egg stages, winter survival is not an issue in the coldest northern regions because all turtles lay their eggs early enough in the summer for them to hatch by fall in most years. However, not all turtle hatchlings leave the nest immediately after hatching, and some species are adapted to spend the entire winter beneath ground and to withstand temperatures that may remain below freezing for weeks. Baby painted turtles (*Chrysemys picta*) of the northern United States and southern Canada typically remain underground over the winter after hatching in the summer or fall, finally emerging the following spring, about a year after the eggs were laid. The hatchlings are able to survive the subfreezing temperatures by producing a chemical that functions like antifreeze in their bodies so that retained water does not freeze and rupture cells. The insulation provided by a few inches of soil is usually adequate to keep temperatures in the nest above freezing during short cold snaps. An additional insulation of snow cover can also effectively maintain nest temperatures above freezing.

Adult turtles avoid freezing temperatures during the winter by selecting hibernation areas that do not remain frozen for extended periods. Terrestrial species, such as tortoises and box turtles (*Terrapene*), rely on soil and vegetation for insulation from the cold. The deep burrows of gopher tortoises (*G. polyphemus*) are well insulated by the soil around them and do not drop below freezing even during the coldest winters throughout their range in the southeastern United States. Box turtles characteristically bury themselves under dead leaves, logs, and other ground litter, including the soil, where they remain insulated from much colder temperatures on the surface. Aquatic turtles that live in rivers or large lakes use the water as the primary insulator from cold-air temperatures, seeking out deeper areas that are unlikely to freeze even during periods of extreme cold. Some species can remain completely submerged in near freezing waters for weeks or months because of their lowered metabolism and by using underwater respiration. Several kinds of turtle that live in seasonal freshwater wetlands, such as those in Carolina bays of the southeastern United States, spend winters on land buried in the surrounding habitat in a similar manner to box turtles.

How do turtles survive droughts?

Turtles that live in deserts are subjected to permanent drought and have developed a suite of traits to survive. In more humid regions of the world, where droughts occur only sporadically or unpredictably, many turtles are adapted to aestivation for long periods. These species can sequester themselves on land hidden beneath ground litter and may remain dormant for extended periods of drought. During aestivation, a turtle's metabolism is lowered considerably so that minimal energy is required and feeding is not necessary. Likewise body fluids are retained and conserved.

What is hibernation?

The classical definition of hibernation is a state of winter dormancy in which animals (traditionally certain mammals) do not eat or drink and during which time the individual's metabolism is lowered and minimal energy is expended. Aestivation, which is similar to hibernation in that an animal lowers its metabolism and remains dormant, usually refers to periods during summer and in some instances to periods of drought in general. Physiologists frequently use the term *brumation* instead of hibernation to refer to ectothermic (cold-blooded) reptiles, such as snakes and turtles, in

Family: Geoemydidae; Species: *Callagur borneoensis;* Common Name: Painted terrapin; Geographic Region: South-east Asia

Family: Geoemydidae; Species: *Callagur borneoensis;* Common Name: Painted terrapin; Geographic Region: South-east Asia

Family: Carettochelyidae; Species: *Carettochelys insculpta*; Common Name: Pig-nosed turtle; Geographic Region:
Australia, New Guinea

Family: Cheloniidae; Species: *Caretta caretta*; Common Name: Loggerhead sea turtle; Geographic Region: Pantropi-
cal and temperate oceans

Family: Chelidae; Species: *Macrochelodina parkeri;* Common Name: Parker's long-necked turtle; Geographic Region: New Guinea

Family: Cheloniidae; Species: *Chelonia mydas;* Common Name: Green sea turtle; Geographic Region: Tropical to temperate seas

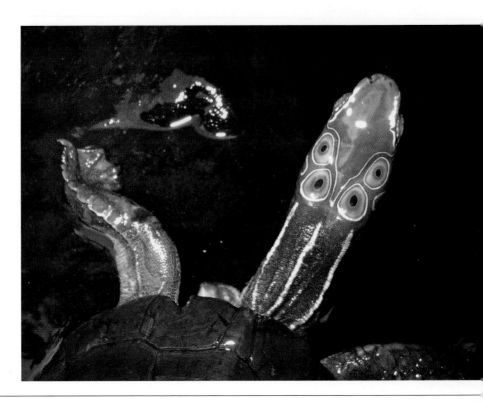

Family: Geoemydidae; Species: *Sacalia quadriocellata*; Common Name: Four-eyed turtle; Geographic Region: South east Asia

Family: Geoemydidae; Species: *Malayemys macrocephala*; Common Name: Malayan snail-eating turtle; Geographic Region: Southeast Asia

Family: Chelidae; Species: *Chelus fimbriata;* Common Name: Matamata; Geographic Region: South America

Family: Chelidae; Species: *Chelus fimbriata;* Common Name: Matamata; Geographic Region: South America

Family: Geoemydidae; Species: *Orlitia borneensis;* Common Name: Malayan giant river turtle; Geographic Region: Southeast Asia

Family: Platysternidae; Species: *Platysternon megacephalum;* Common Name: Big-headed turtle; Geographic Region: Southeast Asia

Family: Emydidae; Species: *Chrysemys picta;* Common Name: Painted turtle; Geographic Region: North America

Family: Emydidae; Species: *Clemmys guttata;* Common Name: Spotted turtle; Geographic Region: North America

Family: Geoemydidae; Species: *Cuora galbinifrons;* Common Name: Indochinese box turtle; Geographic Region: Southeast Asia

Family: Dermatemydidae; Species: *Dermatemys mawii;* Common Name: Central American river turtle; Geographic Region: Mexico, Central America

Family: Emydidae; Species: *Actine-mys marmorata;* Common Name: Pacific pond turtle; Geographic Region: North America

Family: Cheloniidae; Species: *Eret-mochelys imbricata;* Common Name: Hawksbill sea turtle; Geographic Region: Tropical to temperate seas

Leeches are a common parasite on the shell and soft parts of many freshwater turtle species, which they may rid themselves of by basking. Leeches presumably cause little or no permanent damage to the turtle.

Photo © Whit Gibbons

which dormancy is physiologically different from that typical of hibernating mammals, such as bears and ground squirrels.

Do all turtles bask?

Although basking is an obvious behavior of many aquatic species, some turtles almost never bask out of the water, although they may bask aquatically by absorbing the sun's rays below the surface in shallow water or by warming up while floating on the surface. Most species of mud turtles (*Kinosternon*) seldom if ever bask, rather spending their lives prowling along the bottom in search of prey, leaving the water only to hibernate, aestivate, or lay eggs. However, the closely related common musk turtles (*Sternotherus odoratus*) are often found basking along rivers and in swamps but usually in the shade. Individuals will climb up into trees above the water, sometimes several feet above the surface. Perhaps they leave the water to dry out rather than to warm up, possibly to rid themselves of leech parasites that are commonly found on them, sometimes in high numbers. Green sea turtles (*C. mydas*) are known to bask on beaches in Hawaii and in Australia, as well as in the Galápagos Islands, perhaps to raise their body temperature. Some turtle biologists have suggested that unreceptive females bask out of the water to escape the attention of males interested in mating. Other species of marine turtles have been noted to haul ashore and bask at various localities around the world but such behavior is not commonly seen.

Do turtles have enemies?

The enemies of turtles come in many shapes and sizes but have the greatest effect on nearly all species during the egg, hatchling, and early ju-

Giant tortoises living on oceanic islands and having no historical cause for concern for predators often sleep with the neck extended and potentially vulnerable. Photo © Judy Greene

venile stages. Natural predators of turtles include mammals that seek out their eggs during nesting season, including raccoons, opossums, skunks, and foxes. Kingsnakes in the southeastern United States are noted for homing in specifically on nesting aquatic turtles and eating eggs in the nest. Large fish and at least one kind of water snake are documented to have had hatchling turtles in their guts when dissected. Introduced fire ants, which are a scourge to many native species of the Southeast, have been documented to eat eggs and even recent hatchlings still in the nest. Monitor lizards are major predators of turtle eggs and small turtles in many parts of Africa, Asia, and Australia.

Crows and ravens are especially adept as predators on turtle eggs. In areas where aquatic turtles historically nest year after year, crows have been observed to watch females leave the water to nest and then wait patiently for the eggs to be laid after which the crow excavates the nest and eats the eggs. Crows have also been observed preying on hatchlings as they emerge from the nest and as they make their way to the water. Herons, egrets, and other wading birds are potential predators of young freshwater turtles, and alligators, caimans, and crocodiles are a continual source of predation on smaller individuals that venture into open water. Bald eagles have been known to capture diamondback terrapins (*Malaclemys terrapin*) and common musk turtles (*S. odoratus*) and to take them back to their nest to eat them or to feed their young.

Sea turtles' nests are especially vulnerable to mammalian predators such as raccoons or dogs in inhabited areas but also in natural areas where a variety of native predators have been reported to dig up sea turtle nests. According to Meg Hoyle, director of the Learning Through Loggerheads program in South Carolina, feral pigs on the barrier islands of the Atlan-

Turtles: The Animal Answer Guide

Eastern box turtle (*Terrapene carolina*; *top*) and ornate box turtle (*Terrapene ornata*; *bottom*). The high-domed shell of North American box turtles is an effective defense against many biting predators that are unable to crack the shell. Photos © Mike Plummer

tic Coast, aside from causing extensive habitat destruction by rooting and digging, can also prey on sea turtle nests. Sea turtle hatchlings often leave the nest at night and become prey to ghost crabs on the beach, and the stragglers that are still moving across the sand during early morning can become prey of seagulls. Even adult sea turtles have enemies: shark predators will bite off a flipper if given the opportunity, and some of the largest sharks may even be able to crack the shell.

People are the primary enemy of turtles through many direct and indirect activities. When habitats are destroyed or altered (for example, by seawalls) so that nesting areas are no longer suitable, females are forced to lay their eggs in locations that may make them more vulnerable to predators, to flooding, or to high temperatures. These threats would often be avoided if the female could nest where she would have chosen. An unexpected effect of human activity is that hatchling sea turtles are frequently confused by artificial beach lighting and when emerging from the nest at night head in the wrong direction, away from the ocean where they are then killed by predators, on roads, or by desiccation. This has prompted many conservation-minded beach communities to implement education programs like Edisto Beach, South Carolina's "Lights off for Turtles" campaign complete with fines for offenders. Marine turtles in open waters are vulnerable to accidental drowning in nets used in shrimping and in other types of fishing.

Millions of turtles are killed every year on U.S. highways, including many adult turtles that normally would have few natural predators. Turtles of many species are being dramatically reduced in population sizes or even extirpated over large regions because of their popularity as a dietary staple, delicacy, or for falsely presumed health benefits, primarily in China and other Asian countries. As their native species have become increasingly rare, many Asian countries have begun to import large numbers (many thousands) of turtles from as far away as the southeastern United States. Many turtles are sold throughout the world in the pet trade. Although these animals may be well cared for as individuals by a knowledgeable person, those that have been wild caught are forever removed and may never reproduce or help maintain the population that was their source.

Do turtles get sick?

Turtles can get sick from natural situations, and some are especially susceptible in captivity if they are not properly cared for. For some turtles, the stress of captivity is undoubtedly a fundamental cause of problems. Pet turtles are prone to illnesses due to vitamin deficiencies, including vitamin D, which is acquired from ultraviolet light from sunlight. Among the common problems observed are softening of the shell in hard-shelled turtles and eye infections.

A malady commonly observed among wild box turtles is an accumulation of keratin in the inner ear that results in large swellings on one or both sides of the turtle's head. Research has revealed that this is a manifestation of a vitamin A deficiency and occurs with exposure of the turtle to organophosphates found in pesticides and fertilizers commonly used on crops and rights of way. The swelling may prevent the turtle from withdrawing its head and closing its shell, which can interfere with its natural defense from mammalian predators.

Under natural conditions, turtles experience respiratory ailments, such

Fibropapilloma, an unpleasant-looking viral disease, causes external tumors on green sea turtles. Although it is unsightly when present on the neck, eyes, or other body parts, many individuals recover from the ailment. Photo © Cris Hagen

as the upper respiratory tract disease (URTD) prevalent in many populations of desert tortoises (*G. agassizii*) and gopher tortoises (*G. polyphemus*). URTD is best equated in human terms to a form of pneumonia, which can be fatal for some individuals and from which some individuals recover completely. A disease known as fibropapilloma is an ailment found in some species of sea turtles, particularly green sea turtles (*C. mydas*). This disease takes the form of external tumors.

Turtles can presumably contract other diseases that afflict cold-blooded animals, but most are eliminated or their effect is diminished through natural selection that eliminates susceptible individuals that die or whose reproductive output is diminished.

How can you tell if a turtle is sick?

A normal, healthy turtle is not necessarily a picture of high energy and activity, but a sick turtle is readily recognized by people familiar with turtles and their habits. A sick turtle will be listless, stop eating, and may have watery, glazed, or infected eyes or a shell that is unusually soft or even infected with fungus. Some sick turtles will gulp as if attempting to take in air. A common ailment of some pet turtles, especially those that have been held under improper environmental conditions, is shell rot.

Are turtles good for the environment?

Native turtle species within any region serve an important role in ecosystem function as predators that eat other animals and plants, and as prey,

The importance of freshwater turtles in dispersal of aquatic plants is exemplified by the small garden growing in dried mud on the back of a common snapping turtle (*Chelydra serpentina*) that had aestivated alongside a South Carolina wetland during a drought. Photo © Mark Mills

A gopher tortoise basks at the edge of its burrow to which it can readily retreat if threatened by a predator. Burrows are generally the width of the tortoise's carapace length, allowing the individual to turn completely around if it needs to retreat when it is in the burrow. Gopher tortoises are considered keystone species because they can significantly modify the habitat they live in. Photo © Tom Jones

primarily in the egg or juvenile stages, for other species. Gopher tortoises (*G. polyphemus*) and African spurred tortoises (*Centrochelys sulcata*) are keystone species not only because they create habitats used as refuges by hundreds of other animal species but because they also influence vegetation patterns by soil excavation and the creation of microhabitats that enhance growth and productivity for some plants.

Birds have long been recognized as seed dispersers, but other animals, including turtles, provide the same service, often for many species of plants that birds do not eat. Although this phenomenon has been studied in only a few kinds of turtles, it is reasonable to expect that all turtle species that eat the fruiting bodies of plants will deposit viable seeds in other locations,

Turtles: The Animal Answer Guide

thus assisting the distribution of plant species in a region. This can be particularly important after drought or fire has altered the landscape.

One of the most fascinating potential roles of a plant/tortoise interaction was suggested by John Iverson of Earlham College regarding the part the now extinct tortoises that were on the island of Mauritius may have played to enhance the germination of the large seeds of tambalacoque trees. Previous investigators had proposed that the dodo, also native to Mauritius, had played a role in the development of the large seeds of tambalacoque trees in which germination was enhanced by abrasion of the seeds as they passed through the intestinal tract.

Iverson pointed out that the tortoises, which went extinct about the same time as the dodo, were as likely candidates as the bird to have a role in seed germination and development of tambalacoque trees. Population levels of the tambalacoque trees had declined dramatically after the extinction of the dodo and the tortoises, presumably becoming endangered because their natural dispersers of the seeds had disappeared. Not all biologists accept that tambalacoque trees are fully reliant on herbivorous animals that are capable of eating the seeds, whether dodo or tortoise, but this is indeed an intriguing proposition.

Chapter 6

Reproduction and Development

How do turtles reproduce?

Turtle reproduction follows the basic format of most other vertebrates. The mating season for most species in the temperate zones begins in late winter or early spring as temperatures begin to increase, with males seeking out females and engaging in courtship activities, which vary among species. The mating season in tropical areas is more often related to the wet/dry season cycle. Turtle biologists assume that all species have a ritualized courtship process, but courting behavior in the wild has been observed in very few. In some tortoises and snapping turtles (*Chelydra serpentina*), adult males will engage in physical combat with other males to drive them away from their territory, therefore preventing opportunities to mate with local females. Male tortoises will charge one another head-on and butt their shells together. The loser eventually retreats or, in rare instances, is turned over on his back.

Snapping turtle males fight in the water, creating turmoil on the surface of a lake or pond as they claw and bite one another. Observers of snapping turtle combat differ in their interpretations of the details of the process, but the loser usually retreats to another area of larger bodies of water and may even migrate to another lake or stream. Male-male combat probably occurs in other species of turtles in which the sexes are the same size or in which males are larger but evidence is anecdotal and fighting is not well documented. Aggressive behavior of males competing with other males during the mating season does not commonly occur among the many species of turtles in which females are the larger sex.

Complex courtship rituals have been observed between male and female turtles of some species. Adult males of painted (*Chrysemys picta*), slider (*Trachemys scripta*), and some map turtles have elongated foreclaws used in an elaborate courtship behavior called "titillation" in which the male extends his front feet and vibrates his long claws in the water in front of the female. A female interested in mating will follow the male as he swims backward. Many other species of aquatic turtles probably have intricate and complex courtship rituals, but the difficulties of observing a short-term event that occurs underwater for only one or a few days once a year are paramount.

Actual copulation by most or all turtles occurs by the male mounting the carapace of the female and curling his tail under hers until their cloacas are in contact and he can insert his penis. As a general rule, male turtles that mate on land have an indented plastron that accommodates mounting the female so that he is not balancing on her domed shell, a precarious feat at best with a flat plastron. Such indentations are evident in many tortoises and box turtles (*Terrapene*). Species that mate in the water have less obvious concavities on the plastron, as the buoyancy provided by the water allows the male to maintain his position anyway, but the males of some semi-aquatic species, such as the mud turtle (*Kinosternon subrubrum*), are slightly indented. The act of copulation is aided in most species of aquatic turtles by males having appreciably longer tails than females. The male's cloaca is located farther back, which permits him to reach beneath the tail of the female. Male mud turtles (*K. subrubrum*) have a claw on the tip of the tail that is presumably used to provide a better purchase during the mating process. Successful copulation can be completed in a few minutes in some aquatic species under certain situations, whereas the process may last for several hours in some terrestrial box turtles and tortoises.

Do all turtles lay eggs?

Yes. All turtles, like birds and crocodilians, lay eggs. Most fish, snakes, and lizards also lay eggs, but females of some species give live birth. Even two kinds of mammals, the duckbill platypus and spiny anteater, lay eggs. Turtle eggs do not exhibit the variation in color seen in bird eggs; rather, all of them are white or cream colored, but they do vary in shape with some species laying completely spherical eggs while others are elongate. In North America, the largest species, such as snappers (*C. serpentina*), sea turtles, and softshells, lay spherical eggs, but the smaller species lay elongate eggs. However, the smallest flap-shelled turtles (genus *Lissemys*) of Asia lay round eggs, whereas the large mangrove terrapins (*Batagur baska*) of India and Southeast Asia lay elongated eggs making a hard and fast explanation for egg shape elusive. Egg shell texture varies considerably among different families of turtles, ranging from the hard, brittle-shelled eggs of softshell

Turtle eggs do not exhibit the variation in color seen in bird eggs, rather all of them are white or cream colored, but they do vary in shape. Some species lay round eggs while others lay elongate eggs. Photo © Steve Gotte

and mud turtles (*Kinosternon*) to the leathery shell of slider (*T. scripta*) and painted turtles (*C. picta*).

Why do sea turtles lay so many eggs but box turtles lay only a few?

Questions about the number and size of eggs and the number of clutches of eggs turtles lay in a year or in a lifetime by a turtle have intrigued turtle biologists for decades. Over the evolutionary history of any successful species, the number of eggs laid by a female has been a trade-off against the size of the eggs themselves, as larger eggs mean the female can lay fewer with the same amount of resources. Some scientists have proposed that an optimal egg size exists for a species living under a particular set of environmental conditions. Although the concept is debated among theoretical biologists, an unassailable fact is that a turtle egg that is too small may have insufficient yolk to nourish the embryo until hatching and still have yolk reserves available for the newly emerged juvenile to survive until it finds its own food. Producing a smaller-bodied turtle could possibly compromise the survival of the hatchling and its rate of growth to maturity. So, the range in size of box turtle (*Terrapene*) eggs or of any species' eggs

Turtles: The Animal Answer Guide

is a product of natural selection. Another factor in the mix is that the body of a female has limited space to hold eggs; therefore, if the eggs are to be large enough for incubation and nourishment later, then for a box turtle, mud turtle (*Kinosternon*), or any other small-bodied turtle, for example, the number of eggs must be small. Egg size can be constrained in some species by the size of the pelvic opening, the passageway through which each egg must pass to exit the female's body.

Sea turtles, however, which sometimes lay more than a hundred eggs, have plenty of room for storing them. Sea turtle eggs are larger than box turtle eggs, but relative to the body size of the females, they are proportionately smaller but represent a size that has worked over evolutionary time. Although enormous numbers of eggs are produced by a nesting sea turtle, the odds of an individual hatchling surviving to adulthood under natural conditions are extremely low, both because of mortality in the nest and predation on juveniles as they cross the beach and after they reach the sea.

Most species of animals alive today have population sizes that increase in some years and at some locations while other populations decrease in size, but the overall population does not decline enough for the species to go extinct. Interestingly, a relatively stable population size implies that, during its lifetime, an average female will have two (a male and a female) offspring that reach maturity and become a reproducing part of the population. So, although a loggerhead sea turtle (*Caretta caretta*) may lay more than 500 eggs in a year compared with fewer than a dozen by a box turtle, the female's lifetime average egg survivorship to adulthood will be about the same for both species.

How long do female turtles hold eggs in their body?

The length of time that turtles hold shelled eggs in their bodies is known for only a few species. The shortest average period that has been documented is about two weeks based on the time interval between nesting by tagged females of loggerhead (*C. caretta*) and green sea turtles (*Chelonia mydas*). Nat Frazer of the University of Georgia reported an average period of 13 days between nesting by loggerhead sea turtles on Little Cumberland Island, Georgia, although some individuals returned to nest in as few as 8 days. Similar interpretations of two weeks to a month have been determined for slider turtles (*T. scripta*) and eastern mud turtles (*K. subrubrum*) at the Savannah River Ecology Laboratory (SREL) based on x-ray photography (radiography). Turtles that were collected at drift fences with pitfall traps at Ellenton Bay were x-rayed as they left the wetland to nest and x-rayed again when they returned. The full complement of eggs in the

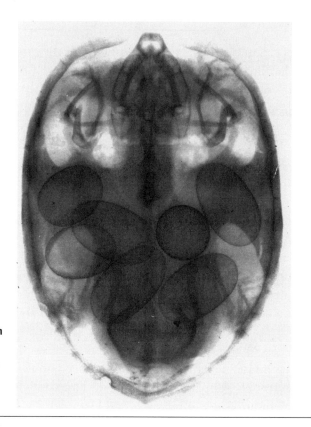

X-ray photography permitted a given clutch of eggs to be identified by their location within the body cavity and juxtaposition to one another.

Photo © Judy Greene

clutch was visible on the day of their departure, and when they reentered the wetland one to several days later, no eggs were visible if the female had nested. On many occasions, the same female again exited the wetland during the same season and again was found to be carrying eggs. In this way, a minimum time period could be determined for how long the particular female had held eggs based on the day on which no eggs were present until the day she was caught again at the drift fence when she returned to lay a second clutch, and then reentered yet again with no eggs present. On the basis of hundreds of captures of female slider and mud turtles with eggs in this study, a range of time for these two species could be established as between two weeks and a month for how long shelled eggs are held.

Some species of turtles have also been known to hold a shelled clutch of eggs for several months before nesting. A dramatic example of female turtles carrying eggs long term is that of the chicken turtles (*Deirochelys reticularia*) studied by Kurt Buhlmann at SREL. X-ray photography permitted a given clutch of eggs to be identified by their location within the body cavity and juxtaposition to each other. Individual females were documented to hold the same clutch of eggs for months without laying them in a nest. One female chicken turtle held a clutch for six months before nesting. The eggs were then monitored, and healthy young hatched after normal incubation,

Turtles: The Animal Answer Guide

Florida cooters are unusual in that a nesting female often digs satellite nests on either side of the main nest cavity and lays one or two eggs in these side nests. Photo © Ken Dodd

putting to rest the concern that eggs held for such a long time in the female's body, for this particular species at least, would no longer be viable.

Where do turtles lay their eggs?

Most turtles lay their eggs on land, usually in areas with adequate sunlight to keep the soil warm and sufficient moisture to keep the eggs from drying out. Marine species lay their eggs on beaches, and river turtles often dig their nests on sandbars. Gopher tortoises (*Gopherus polyphemus*) of the southeastern United States frequently lay their eggs in the soil at the mouth of their burrow. Because nesting is a critical part of the life cycle that is not to be denied, turtles that live in habitats that have been developed as suburban areas for humans will leave lakes and ponds to dig their nests on lawns, golf courses, road shoulders, and in flower gardens. Some species travel long distances through the ocean, up or down rivers, or overland to select favorable nesting sites.

One species of the snake-necked turtles of the Australian tropics, the northern snake-necked turtle (*Macrochelodina rugosa*), is the only turtle in the world known to lay its eggs underwater. Until an Australian biologist named Rod Kennett reported his research in the 1980s, scientists were not

The signature tracks in the sand of a female sea turtle that has nested come from and lead back to the ocean. Photo © Kurt Buhlmann

aware that any species of turtle did this, although the natives of the region had apparently known for centuries that these long-necked turtles nested underwater. Kennett pursued the nesting habits of the turtles after discussions with the Aboriginal people of northern Australia where the wetlands known as billabongs alternate seasonally between wet and dry. He confirmed the unusual nesting behavior of the species through the use of radiotelemetry by placing transmitters inside the oviducts of female long-necked turtles that were carrying eggs. Thus, the transmitter that was sending out a constant signal was deposited in the spot where the eggs were laid. By tracking the transmitters, Rod was able to locate the nesting sites, which were indeed beneath the water. He further discovered that the eggs hatch after the waters of the billabong recede during the dry season.

Turtles: The Animal Answer Guide

Diamondback terrapins frequently dig their nests in the sand dunes between the ocean and the brackish tidal creeks where they live. A female may return to nest in the same general area in different years.

Photo © Whit Gibbons

Does a turtle nest at the same time and in the same place every year?

Some species of turtles return to a general nesting area year after year and may select a nesting spot only a short distance from or in the same spot as in previous years. However, turtle nests are not like those of some birds, such as eagles and hawks, which return to the same nest year after year. Once hatchlings leave a turtle nest, the eggshells gradually disintegrate and the soil returns to the type characteristic of the area. Sea turtles such as loggerheads (*C. caretta*) and green sea turtles (*C. mydas*) that may lay several clutches of eggs in a single season have been observed to return to the same beach as many as a half dozen times to lay eggs during the nesting season but not in the same spot.

Individuals of some turtle species that have been studied have been shown to select particular environmental characteristics for egg laying. For example, during studies with eastern mud turtles (*K. subrubrum*) at

Ellenton Bay in South Carolina, University of Georgia graduate student Vincent Burke noted that particular females, which could depart from an oval-shaped wetland in any direction, would almost always choose the same direction, with different females consistently selecting a specific area. Almost the entire perimeter of the wetland seemed suitable for nesting from the human point of view, but the female turtles evidently did not agree and did not exit randomly. Instead they made their way to nest in their personally favored areas.

Do turtles nest only one time per year?

Females of some species of turtles never nest more than once a year whereas individuals of other species commonly nest at least twice a year and sometimes more. Nesting frequency has no apparent relationship to body size of the species that have been studied.

Studies on loggerhead (*C. caretta*) and green sea turtles (*C. mydas*) on the Atlantic coasts of North America and Central America have provided the best long-term records of nesting frequency of individuals and seasonal patterns within species of sea turtles. Individual females of the enormous loggerhead and green sea turtles nest multiple times, averaging three to four times, in a year. Reports have been made of green turtle and loggerhead females laying more than six times in a year. After laying eggs on a beach, the female returns to the ocean for about two weeks and then nests again. However, individual loggerhead and green sea turtles do not usually nest in consecutive years but skip one or more years before nesting again.

Probably the most intensively studied freshwater turtle populations in which egg-laying patterns have been recorded for decades are from research programs by herpetologists at SREL and those conducted by Dr. Justin Congdon at the University of Michigan's George Reserve in southern Michigan. Hundreds of observations of large common snapping turtles (*C. serpentina*) in South Carolina, Michigan, and elsewhere have never documented a second clutch by a female in a year. Yet at these same sites, females of painted turtles (*C. picta*) in Michigan and slider turtles (*T. scripta*) and eastern mud turtles (*K. subrubrum*) in South Carolina have been observed on numerous occasions to lay two or three clutches in the same year.

Within a turtle species the nesting season is fairly well defined for most that have been studied. Loggerhead sea turtles (*C. caretta*) along the Atlantic Coast of the southeastern United States usually begin nesting in late May through early August, with the peak intensity in June and July. Slider turtles begin nesting as early as April and May and continue through June into July, with the last ones to nest being ones that have laid one or more

Thread trailers designed to be placed on the back of a turtle leaving an aquatic habitat has allowed turtle biologists to determine where nesting and hibernation sites are. Photo © Peri Mason

clutches earlier in the season. Painted turtles in Michigan begin nesting in May and continue into June. The timing varies by a few days from year to year, depending on temperature and rainfall. Snapping turtles appear to have a short nesting period of no more than a month every year. Although the females of a species at a particular habitat lay eggs around the same time, the initiation and length of the season vary each year in response to spring temperatures and rainfall.

One of the most complex, almost mysterious, seasonal nesting patterns yet observed for any turtle is that of the chicken turtle (*D. reticularia*), a species studied for several years by Dr. Kurt Buhlmann and his colleagues at the University of Georgia's SREL. Chicken turtles in South Carolina begin laying eggs as early as August and may nest in any month until April; most of the nesting period is in fall and winter. A female may actually nest in the fall and then again later in the winter. The seasonal timing of nesting is in contrast to the nesting periods of most other North American species in which the nesting season is primarily spring and summer.

How many eggs do turtles lay?

The number of eggs that turtles lay during a single nesting bout, a season, and a lifetime vary considerably among species. The greatest numbers of eggs laid by a female at a single time and in a single nesting season are those of the sea turtles. Karen Bjorndal of the Center for Sea Turtle Research at the University of Florida used Archie Carr's 2,500 nesting records of green sea turtles (*C. mydas*) taken over a 30-year period in Costa Rica and found the average number of eggs in a clutch to be 112. The highest recorded was 219. Nat Frazer and Jim Richardson of the Univer-

Snapping turtles lay an average of about 30 round eggs in the spring from which the babies emerge in late summer. Photo © David Scott

sity of Georgia found that average clutch sizes of loggerhead sea turtles (*C. caretta*) on Little Cumberland Island, Georgia, were even higher, averaging almost 120 based on nesting records over 19 years. They found the highest average clutch size in a single year to be more than 127 eggs and the lowest 114.

Individual female loggerhead and green sea turtles commonly nest as many as four or more times during a nesting season. Thus, the total number of eggs laid on average by a single female may exceed 400, and the maximum number could be as high as 700 for some. The lifetime expectation for these high fecundity turtles requires additional calculations of the annual frequency at which individual females nest and the reproductive longevity of the turtles. Females of loggerhead and green sea turtles characteristically wait at least one and sometimes more years between nesting bouts, so their lifetime egg output would be half or less of what they produce in a year times their reproductive life expectancy. Assuming that sea turtles live for at least ten years after reaching maturity (a conservative estimate) and average 400 eggs every other year, an average female would lay more than 2,000 eggs during her lifetime.

A general rule is that the number of eggs laid by a female increases with body size. Thus, larger species of turtles are likely to lay more eggs than smaller ones, and larger females within a species are more likely to lay more eggs in a clutch than smaller females. Records for 73 slider turtle (*T. scripta*) clutches from Ellenton Bay, South Carolina, demonstrated that the average clutch size was about six eggs; however, the average clutch size of slider turtles from the Par Pond Reservoir, where the individuals are much larger than those at Ellenton Bay, was around 10 eggs. Likewise, the

Turtles: The Animal Answer Guide

average clutch size of 143 Ellenton Bay mud turtles (*K. subrubrum*), which are much smaller than any adult slider turtles, was only three eggs, and the maximum was only five. The average clutch size of the largest freshwater turtle species on the Savannah River Site in South Carolina, the snapping turtle (*C. serpentina*), was 29 based on a sample size of 37 clutches. The lowest clutch size was 13 eggs and the highest was 55. In a study of snapping turtles in southern Michigan over a six-year period, Justin Congdon found the average of 68 recorded clutch sizes to be about 30 eggs, the minimum was 12, and the maximum was 41. A clutch size of 104 eggs was recorded for a snapping turtle from Nebraska, indicating the range in variability that can occur among individuals, even within a single species.

Are all hatchlings in a turtle nest full siblings?

A female turtle can lay a clutch of eggs fertilized by more than one male, which would result in some of the offspring being half-siblings. Tracey Tuberville used DNA analyses to determine the male parentage of gopher tortoise (*G. polyphemus*) hatchlings on St. Catherines Island, Georgia. She found that females were just as likely to have clutches fathered by two males as they were to have clutches fathered by a single male. At least one female was suspected of having a clutch in which eggs had been fertilized by as many as three males. Presumably females of many other turtle species are fertilized by more than one male during the mating season and are capable of laying a clutch of eggs in which the offspring have two or more fathers.

How is the sex of a turtle determined?

Whether an egg will produce a male or female is determined genetically at the time of conception in some species of turtles, which is true with most vertebrates. However, in a widespread phenomenon known as temperature sex determination (TSD; also referred to as environmental sex determination, or ESD), sex is determined by temperatures of the nest during the incubation period. Softshell turtles that have been investigated are assumed by researchers to have sex chromosomes, and possibly other species do as well, because the sex ratio of males to females is 1:1 no matter what the temperature of the nest, indicating that their sex is determined genetically. But in most species of turtles that have been examined, sex is determined by the temperature of the nest during a particular portion of the incubation period. The temperatures before or after are less influential.

The simplest explanation of TSD is that, when turtle eggs are incubated at warm temperatures, females are produced and when temperatures

Ornate box turtles mate on land and can become attached so that the female actually drags the male along behind her. Females of several species of turtles have been known to mate with more than one male so that hatchlings from the same clutch of eggs may have different fathers. Photo © Mike Plummer

are cooler, males are produced. As is often the case with scientific studies, as the details of a biological phenomenon are investigated, numerous variables are found that affect the phenomenon. For example, one study with snapping turtles (*C. serpentina*) showed that females were produced at the highest incubation temperatures and the lowest ones, while mostly males were produced at intermediate temperatures. Several natural variables can also influence the sex ratio within a nest. Thus, a sea turtle nest laid on an open beach exposed to sun may produce different ratios of males to females based on how deep the nest is, how many eggs are at the bottom of the nest where temperatures are significantly cooler than those at the surface, how many cloudy or rainy days there were during the time of incubation, and other factors that would influence the temperature of the soil surrounding the eggs.

Temperature or environmental sex determination is a complex biological phenomenon that occurs in many species of egg-laying and, possibly, even live-bearing reptiles, and full understanding of all ramifications of the process are still being sorted out.

Do turtles care for their young?

No species of turtle is known to protect or care for its young throughout the incubation period or once the hatchlings have left the nest; how-

Turtles: The Animal Answer Guide

The position of the yolk scar of a diamondback terrapin that recently hatched and emerged from the egg is evident with the accumulation of sand in the center of the plastron. Turtles exhibit parental care before laying their eggs by providing their young with a several-month supply of nourishment in the form of yolk.

Photo © Tom Luhring

ever, the female Asian brown tortoise (*Manouria emys*) has been reported to remain in the vicinity of the aboveground nest, in some instances, and will physically ward off potential predators. Among the natural predators of the eggs are monitor lizards, which the female tortoise has been reported to have no reservations about charging and attempting to bite. The female tortoise will also move aggressively toward a human who approaches the nest site. The behavior toward humans has been observed in animals in captivity, but whether this defensive behavior is widespread in the wild has not been verified. Chuck Shaffer observed a female in the wild that left the nesting site soon after depositing the eggs. One needs to remember that turtles are also individuals, and, like humans, some may be better mothers than others.

Gopher tortoises (*G. polyphemus*), which often lay their eggs in the entrance of their burrows, have been reported to charge upward from the burrow in which eggs are present, presumably in an aggressive attack to drive away a potential predator. Whether such behavior would continue throughout incubation or later when hatchlings are present is unknown. Although similar nest guarding behavior may occur among other turtles, the phenomenon either has not been observed or has not been reported in the scientific literature.

Although turtles are for the most part seemingly uncaring for their eggs or young after birth, they actually provide extensive prenatal parental care, which occurs not only before an embryo hatches but also before fertilization of the egg, a phenomenon noted in the scientific literature by Justin Congdon. The females store large quantities of fat before ovulation of the eggs so that when the female lays her clutch, each individual embryo is provisioned with enough nutrients to last for several weeks or even months without needing to eat. Chicken turtles (*D. reticularia*) have been documented by Kurt Buhlmann to go 18 months between the time the eggs were laid until the hatchlings enter the aquatic habitat and begin feeding. During that time, the embryo and hatchling lived off the food resources provided solely from the yolk material provided by the female. While turtles do not care for the young during incubation and after hatching, they successfully prepare them for a good start in life by providing stored energy reserves.

How fast do turtles grow?

Compared with mammals and birds, turtles grow very slowly. Some species grow faster than others, and the rate of growth within the same species can vary dramatically because of environmental conditions. Growth rates of slider turtles (*T. scripta*) and painted turtles (*C. picta*) have been studied more extensively than most other species, and in both of these, examples have been documented in which growth rates in body length can be significantly faster in one population compared with another only a few miles away. Slider turtles from a well-studied population in Ellenton Bay, South Carolina, a natural freshwater wetland, grew at a rate of about half an inch per year as juveniles.

Individuals in a population at Par Pond, a 2,600-acre reservoir that received thermally elevated waters from the cooling systems of nuclear production reactors, grew at an average rate of twice that or more. The body weight increases of turtles in the Par Pond population were also significantly greater than those of any natural populations in the region. Some of the fastest growth rates in turtles have been those reported for captive tortoises and small leatherback sea turtles (*Dermochelys coriacea*) that have been raised on unrestricted diets. However, even the most rapid growth rates in length or weight recorded for turtles in the wild are not close to the rates observed in mammals or birds of comparable size.

A characteristic feature of turtles as well as vertebrates in general is that growth rates are most rapid in the juvenile stages and decrease significantly or almost entirely once maturity is reached. Some turtles that have been studied carefully have indeterminate growth as adults, which means they have the capability of continuing to grow throughout their lives although

New growth during the year is often apparent in individuals because older parts of the shell are stained, as seen in the light-colored areas of this painted turtle from a habitat high in iron content. Photo © David Scott

at a diminishing growth rate. By comparing measurements of individual mud turtles (*K. subrubrum*) that had been studied during a research program at Ellenton Bay on the Savannah River Site over almost 40 years, Ria Tsaliagos of SREL determined that when environmental conditions were right, individuals were able to continue adding to their length several years after they had reached maturity, sometimes when they had not grown for several years.

How can you tell the age of a turtle?

Compared with most other species of animals, turtles live a long time, with many kinds having been documented with certainty to live for decades, both in the wild and in captivity. The age of many species of hard-shelled freshwater turtles and tortoises can be estimated accurately up to seven or eight years by counting the number of growth annuli, which are grooves on each of the plates (scutes) of the shell. These shallow indentations look somewhat similar to those used to estimate the age of a tree that has been cut down, and, as with tree rings, each usually represents a year's growth. During periods of inactivity, such as winter, a turtle's scutes grow slowly or not at all, and the section of the scute differs in appearance from the spring and summer periods of rapid growth.

Caution is necessary when a person uses the annulus technique to age turtles. For example, some turtles, such as the mud turtles (*Kinosternon*) of

The age of many species of hard-shelled freshwater turtles and tortoises can be estimated accurately up to seven or eight years by counting the number of growth annuli, which are grooves on each of the plates (scutes) of the shell. SREL file photo

North America have reached maturity by the time they are six or seven years old, and their annual growth slows considerably. Thus, although growth rings are produced after maturity, annual growth is minimal, and the rings become tightly compressed, making it difficult to differentiate between them. Rings can also be an inaccurate measure of age in turtles in which the shell has become worn and smooth, and the rings become faint or disappear completely. Environmental conditions can also cause turtles to develop "false annuli." For example, in a year that has lots of available food early in the year, is then interrupted by extreme drought, and then has another period of rainfall and good food availability late in the year, the turtle may show a false, narrow annulus in the middle of that year.

Nonetheless, in some turtle species, fairly reliable estimates of age can be made even after 10 or 15 years or more. Although the count may not be suitably precise for scientific studies, when children count 18 to 20 growth rings on the scutes of a common box turtle (*Terrapene carolina*), ornate box turtle (*Terrapene ornata*), or tortoise, they can be relatively certain that they have an animal that is much older than they are.

Turtles: The Animal Answer Guide

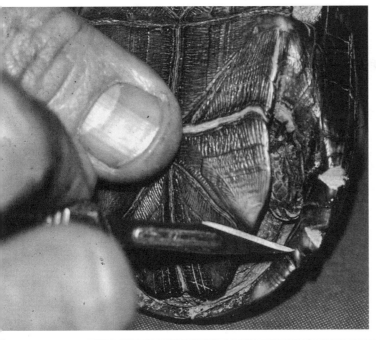

Shell marking is a technique scientists use to determine how long turtles live. The tip of the scalpel points to a permanent mark carefully made with a scalpel when the mud turtle was a newborn hatchling seven years earlier. The mark is still visible as a rounded scar that permits individual turtles to be identified anytime they are captured, even years later, indicating that the shell marking technique is effective even with hatchling turtles. New marks have been made on this individual with a file to ensure that the individual can continue to be identified for years to come. Photo © Judy Greene

How long do turtles live?

Turtles live a long time, and individuals within natural populations are known to live longer than most other animals of any sort. However, despite claims of a centenarian turtle at one place or another, true documentation is not always forthcoming. People like to be able to say they have a record, and "oh my" statistics are always gratifying to present. But some of the records reported, even by turtle biologists, should be viewed as suspect unless a clear chain of custody exists for the specimen, especially when someone claims that someone else's grandfather caught the turtle as a boy and put his initials on the shell. Not to say turtles do not live a long time and that some individuals have not lived more than a century, but records must be carefully considered.

Turtle biologists can sometimes combine the age estimates from growth rings with another technique known as mark-recapture to determine that field-caught turtles are much older than could be determined by annuli alone. The marginal scutes of hard-shelled turtles can be readily marked with a file or drill so that individuals can be recognized many years later. In long-term studies with painted turtles (*C. picta*), eastern mud turtles (*K. subrubrum*), and slider turtles (*T. scripta*), individuals that were initially estimated to be 6–10 years old and then recaptured as much as 20–25 years later were clearly more than a quarter of a century old, and some known-age individuals were more than 30.

Two long-term studies on turtles in southern Michigan, one at Sher-

Reproduction and Development

Among the most valuable techniques used in population studies to determine longevity of turtles is marking animals for future recognition of individuals. The marginal scutes of hard-shelled turtles can be readily marked with a file or drill so that individuals can be recognized many years later. AEC, which stands for Atomic Energy Commission, on the towel indicates that this slider is being drilled with an individual code in the 1960s. The individual turtle would later be caught in the 1980s, more than 20 years after being marked. SREL file photo

riff's Marsh associated with Michigan State University's Kellogg Biological Station and the other on the University of Michigan's George Reserve, give unquestioned evidence that some species can live a long time in the wild. An adult Blanding's turtle (*Emydoidea blandingii*) is at least 14 or more years old, and many adults were captured and individually marked in the 1950s on the George Reserve and in the 1960s in Sherriff's Marsh. In the early 2000s, numerous individual Blanding's turtles were captured at both locations that had been first captured as adults at least 40 years earlier. Some were speculated to have been more than 25 years old when first captured, so these would be more than 60 years old.

The oldest turtle of known age to lay eggs in the wild was a Blanding's turtle in Minnesota that was documented by researchers to be more than 75 years old. Some of the greatest age records for turtles have come from animals kept in captivity, primarily from zoo records or from carefully documented records by pet owners. Some reliable records are of African spurred tortoises (*Centrochelys sulcata*) living more than 50 years, Aldabra tortoises (*Aldabrachelys dussumieri*) more than 60 years, and a common European pond turtle (*Emys orbicularis*) and an alligator snapping turtle (*Macrochelys temminckii*) more than 70 years. Claims of box turtles (*Terrapene*) and tortoises living more than a century have been made for many years by many individuals and are difficult to refute or to confirm, but some of the giant tortoises clearly seem capable of extended longevity. Matt Aresco was

Turtles: The Animal Answer Guide

given a male gopher tortoise (*G. polyphemus*) from Florida that someone found in 1964 when it was a small juvenile. The adult was still alive and healthy in 2009, making it at least 45 years old. According to Kim Lovich in the Curator's Department at the Zoological Society of San Diego, several of the giant Galápagos tortoises in their care have notable longevity records, of which 10 acquired in 1928 were still alive in 2008, 80 years later! Scott Pfaff, herpetological curator at Riverbanks Zoo in Columbia, South Carolina, also reports that three of their Galápagos tortoises still alive in 2009 were received as adults in 1928. Other assertions of greater longevity exist, but verifying the records is often difficult.

Chapter 7

Foods and Feeding

What do turtles eat?

Collectively, turtles eat almost any kind of organic matter, plant or animal, dead or alive. Some species are generalists that eat both plant and animal material. Common box turtles (*Terrapene carolina*), for example, will eat berries, fruit, or mushrooms when they are available, as well as insects or earthworms they find on the forest floor, or sometimes even dead animal matter. As many people who have kept box turtles for pets will attest, they also will eat dry or canned dog food. Slider turtles (*Trachemys scripta*) are especially notorious for eating almost any plant or animal food in the aquatic environment, a possible reason for red-eared sliders' success as an introduced species that can thrive in many parts of the world.

Some species have a wide array of diet choices but are primarily herbivorous, as is true of most tortoises, which graze on a variety of plant species. These slowly lumbering giants of the turtle world would certainly be unable to capture most prey by pursuit but will occasionally feed on animal carcasses when they encounter them. Other species, such as most of the softshell turtles throughout the world, are carnivorous but without a dependency on any particular type of animal. But even primarily carnivorous species may occasionally resort to eating plant material as evidenced by a report of alligator snapping turtles (*Macrochelys temminckii*) consuming large quantities of acorns.

Female map turtles (*Graptemys*) are among specialists that focus their diets on particular food items by eating mollusks that live in rivers. A primary food item of leatherback sea turtles (*Dermochelys coriacea*) is jellyfish, which unfortunately can lead to their mistaking plastic bags floating in the ocean as their favored food, leading to intestinal blockage and subsequent death. The aquatic twist-necked turtle (*Platemys platycephala*) of South America has been reported to favor amphibian eggs but will also eat other animal prey. Tony Tucker and Nancy Fitzsimmons of Savannah River Ecology Laboratory (SREL) found that the salt marsh periwinkle, a small snail, constituted almost 80% of the diet of diamondback terrapins (*Malaclemys terrapin*) in the salt marsh at Kiawah Island, South Carolina. The Mekong snail-eating turtle (*Malayemys subtrijuga*) will eat other invertebrates and fish but shows a strong preference for snails and other mollusks.

Slider turtles are an example of how food preferences can change in a species as they grow from juvenile to adult and from a few ounces to several pounds or if they are presented with different feeding opportunities. A study by David B. Clark at SREL revealed that young slider turtles were carnivorous or insectivorous, whereas adults in the same lake were herbivorous. As a student at the University of Georgia, Robert Parmenter followed up with a study that further demonstrated that while slider turtles were herbivorous as adults in some lakes, they were actually opportunistic carnivores, preferring fish or animal food but able to survive on a strict vegetarian diet. These findings were further amplified by Harold Avery at SREL who conducted laboratory studies that documented that higher-protein diets did indeed enhance the growth processes of juvenile slider turtles. His studies also determined that 1–2-year-old juveniles could not maintain growth and actually lost weight and shrank in size, when provided a strict vegetarian diet.

Do turtles chew their food?

Turtles do not have teeth and do not chew food in the conventional manner, although map turtles (*Graptemys*), musk turtles (*Sternotherus*), and diamondback terrapins (*M. terrapin*) crush and grind the shells of mollusks as they eat. Turtles that eat vegetation bite off edible chunks and swallow one bite before taking another. Likewise, scavengers bite off chunks of meat and swallow them whole. The matamata (*Chelus fimbriata*) consumes its aquatic prey of live fish or invertebrates by a unique powerful suction process. Excess water and debris are expelled, and then the turtle swallows the animal whole. Many kinds of aquatic and terrestrial turtles use their front feet to manipulate their food during feeding.

How do turtles find food?

Turtles find their preferred food by searching visually or by smell. Herbivorous species such as river cooters (*Pseudemys concinna*) and green sea

Sexual dimorphism is apparent in some turtles, such as diamondback terrapins, in which adult females have disproportionately larger heads and jaws than males. The broad crushing surfaces of the jaw allow females to eat larger mollusks than do the males. SREL file photo

turtles (*Chelonia mydas*) seek out patches of vegetation they are able to find first by sight and later by remembering where they fed at an earlier time. Some aquatic turtles can detect the smell of dead fish or other animals and follow the scent to a meal. The wood turtle (*Glyptemys insculpta*) of the northeastern United States finds its food by searching for berries or small invertebrates that it can catch. Wood turtles also have a highly unusual and interesting behavior in which they drum their front feet and thump their shell on the ground in moist soils, which brings earthworms to the surface that the turtle quickly grabs and eats.

Some species can be considered true sit-and-wait predators in that their food actually finds them as they patiently wait in one spot. Prime examples of this behavior are the alligator snapping turtle (*M. temminckii*) with its wormlike tongue appendage that is used as a lure while the turtle lies underwater with its mouth open; the bizarre matamata of South America that lies camouflaged on the bottom looking like a piece of underwater debris and then quickly expanding its throat when a fish swims nearby, rapidly drawing the hapless prey into its mouth; and some of the softshell turtles that conceal themselves beneath sand or silt with only their nose and eyes exposed until they have an opportunity to make a rapid strike at a passing fish or invertebrate.

Are any turtles scavengers?

Many species of turtles are opportunistic scavengers that will eat dead animal material. Dead fish in lakes are readily consumed by several species of turtles, including painted turtles (*Chrysemys picta*) and slider turtles (*T. scripta*), which are characteristically herbivorous as adults. Turtles will also

Turtles: The Animal Answer Guide

Among the many means by which turtles capture prey, the wriggling, wormlike appendage in the mouth of an alligator snapping turtle is one of the most effective for luring fish and other aquatic prey within striking distance. Photo © Cris Hagen

eat dead animals on land. An unusual observation was made by Michael Gibbons of Aiken, South Carolina, who watched an adult gopher tortoise (*Gopherus polyphemus*) eating a road-killed armadillo in southern Georgia. When a car approached, the tortoise retreated to the road shoulder; after the car passed, the tortoise immediately returned to the carcass and continued feeding. Most herbivorous turtle species will eat animal material if given an opportunity. While tortoises are way too slow to run down and capture most active prey, scavenging is an effective means of obtaining a high-protein meal. Scavenging is also safer than attacking prey that could fight back by resorting to defensive behaviors that could injure the turtle.

How do turtles eat hard-shelled animals?

Many turtles eat hard-shelled animals, which include crawfish and crabs, clams and snails, and even other turtles. Their sharp-edged jaws and strong jaw muscles equip them to easily crush many hard-shelled crustaceans. Some species of turtles that specialize in feeding on mollusks have large, flattened surfaces on their jaws and can actually crack and crush the shell of many snails and clams. Nonetheless, although diamondback terrapins (*M. terrapin*) have powerful jaws and have been shown to eat mostly salt marsh periwinkle snails at Kiawah Island, South Carolina, further study by Tony Tucker and Rebecca Yeomans of SREL showed that mud snails were not included in the terrapin's diet.

Pressure tests demonstrated that mud snails have extremely tough shells with more than two to three times the resistance to crushing than periwinkles; therefore, a terrapin may be unable to crack them. Other reasons than shell thickness are possible explanations for why the terrapins do not consume mud snails, which are very abundant in the habitat, but the mechanical inefficiency or impossibility of processing mud snails as a food source appears to be the simplest explanation. Although no turtle specializes in eating other turtles, the way some snakes, mammals, and birds specialize in eating other species of their kind, a large alligator snapping turtle (*M. temminckii*), the giant musk turtles (genus *Staurotypus*) of Mexico and Central America, and several other turtle species would not hesitate to eat a smaller turtle.

Do turtles store their food?

When turtles eat food, whether plant or animal, they consume what is available at the time, but they do not store food to eat later. Even gopher tortoises (*G. polyphemus*), which build large, underground tunnels, do not bring food into them to snack on later.

Although turtles do not cache food like many mammals and birds do, they store food internally in the form of lipids, or fat. Large, orange fat globules are stored inside the abdominal region around the stomach, intestines, and other organs. This stored energy can be metabolized during periods when food is scarce or unavailable and is available to females that are producing eggs that require a large yolk, which is made up of fat and other added nutrients.

Chapter 8

Turtles and Humans

Do turtles make good pets?

Considerations for whether a turtle will make a good pet include such factors as the species of turtle and its disposition, health, age, and size when introduced to captivity. Even under the best circumstances, a turtle can only be considered a good pet if it has an owner who will properly care for it and who is not looking for an animal that is cuddly, sociable, or fast. Most individual turtles are easy to feed, easy to clean up after, and do not become a nuisance by barking, scratching furniture, or escaping from a cage. Small turtles can be kept inside in an aquarium or other container. Terrestrial species such as box turtles (*Terrapene*) and tortoises can be kept in an outdoor enclosure or in a backyard that is properly fenced to prevent escape. Tony Mills, a herpetologist and naturalist with the Low Country Institute in coastal South Carolina, modified a stall of a horse barn to house two orphaned African spurred tortoises (*Centrochelys sulcata*) that he was given. He installed commercial reptile warming strips, made a tortoise entrance designed much like a dog door, and fenced a small grassy paddock so that they could go in and out at will. Such efforts may be necessary to properly care for some species of turtles with special requirements.

How do you take care of a pet turtle?

An important and often overlooked consideration in taking care of a pet turtle is knowing what you will end up with. People often start with a cute baby the size of a silver dollar that is easy to handle and end up with an animal that is larger than a suitcase and unmanageable. A prime example is

Box turtles make excellent pets if they are well cared for and usually can be safely handled by children. As with all wild animals, a pet owner should check laws and regulations to be certain that the animal can be legally kept as a pet. Learning proper care for any pet is also an important consideration. Photo © Whit Gibbons

the popular African spurred tortoises (*C. sulcata*) sold in the pet trade and enjoyed by many. A person unfamiliar with the species may unwittingly buy a 2-inch-long, 2-ounce hatchling that can live in a shoebox and later end up with a 2-foot-long tortoise, weighing more than a hundred pounds, roaming the backyard. Plus, it might live for fifty years, leaving the owner with quite a long-term responsibility. So the first step in taking care of a pet turtle is to know what you are getting now and what you will have five years later (or whenever it grows up). Another consideration is that species such as red-eared sliders that are commonly marketed as pets are too common, making it difficult to find homes for an unwanted, overgrown pet.

Information on proper pet care and treatment of pet turtles is easily available from popular books on turtle care and on reputable Web sites. Even red-eared slider turtles (*Trachemys scripta elegans*), the most common turtle in the pet trade, will outgrow the small bowl in which they can be housed as hatchlings. Many people who are concerned with the welfare of their pet turtle and do a good job with proper feeding and health care are still dismayed when their tiny slider turtle becomes the size of a cigar box, creates smelly waste, and requires frequent care and cleaning.

Legal issues must also be considered in keeping a pet turtle. In the United States, federal and state laws and regulations must be observed. For example, more than a half-dozen species were on the U.S. endangered species list at the turn of the twenty-first century, and while their prospects in the wild are unlikely to improve, anyone who is found keeping one as a pet could be prosecuted. Some states prohibit the removal of gopher tortoises

(*Gopherus polyphemus*) and box turtles (*Terrapene*) from the wild, and quota limits are set on other species such as diamondback terrapins (*Malaclemys terrapin*). Wildlife legislation is a dynamic process. The rules and regulations in a particular region could change, but it is the pet owner's responsibility to determine which rules apply to the species of turtles in the jurisdiction in which the pet owner lives.

Cris Hagen of the Savannah River Ecology Laboratory has personally cared for about half of the world's species of turtles and tortoises and has assisted with the care of more than 200 species of turtles from around the world that have been maintained in captivity by others. These include many rare and endangered species. Hagen offers several suggestions about the care of captive turtles. Some key points to consider for any pet turtle is that an adequate temperature range must be maintained and that it must be properly fed a customized and appropriate diet for the species. For example, some tortoises from very dry climates need to be fed dry vegetation diets with very low water content, even though they can develop a taste for juicy fruits and vegetables, such as melons, tomatoes, and grapes. The diets of all species may need to include supplemental vitamins and minerals, such as calcium for proper growth and shell development. Pet turtles must also receive adequate sunlight or an artificial ultraviolet equivalent. The correct balance of water and basking sites for aquatic turtles and shelters for terrestrial species are essential. Last, anyone deciding to keep a turtle for a pet permanently should be aware that most of those that receive proper care may live 20 to 30 years or more.

Are turtles dangerous?

No turtle is dangerous to people except when the animal has been placed in a situation in which it is threatened and attempts to protect itself by biting and, in the case of snapping turtles (*Chelydra serpentina*), also scratching with sharp, powerful claws. Almost any turtle will attempt to escape when confronted by a human. Even an enormous loggerhead sea turtle (*Caretta caretta*) emerging from the sea will retreat back into the ocean if a person approaches, and a snapping turtle prowling in shallow water alongshore will head for deeper water when approached. Almost without exception, no turtle is dangerous to anybody unless the person creates a situation in which the turtle feels threatened and must defend itself.

Do turtles feel pain?

Turtles are able to feel pain if they are injured, although their response differs from the response of a human, dog, or cat because a turtle will make no sound other than a stoic hiss. Evidence of pain, for example, is obvious when a turtle quickly withdraws a limb that is injured. Despite the quiet and reserved nature of turtles, it is reasonably assumed by veterinarians, biologists, and nonprofessional turtle appreciators that physical injury is as painful to them as it would be to humans and many other animals.

What should I do if I find an injured turtle?

The type of injury and the location of the turtle are both important in deciding on a treatment for an injured turtle. If the injury appears to be natural, such as shell or limb injury caused by a predator such as a raccoon or fox, and the turtle is in a natural area, leaving it alone is probably the best approach. But many injured turtles have cracked shells due to human interference, such as being run over by a car, hit by a boat propeller, or chopped by a lawnmower. The triage approach of categorizing the injury as minor, intermediate, or major is probably best, especially if any treatment of the turtle's injury would require taking the turtle a long distance from the site.

Like most of their reptilian relatives, turtles are tough, and a minor, non-life-threatening injury would require no treatment, and the turtle can simply be left alone or released where it is found. A minor injury would include a lacerated leg or a scratched or even slightly cracked or chipped shell, but with no organs being exposed. Turtles have remarkable healing abilities and should recover on their own from such injuries. Even when a turtle loses all or part of a limb to a predator, the wound usually heals without treatment and the turtle continues to use what is left of a limb in swimming, digging, or walking motions as if it were complete and fully functional. Turtle biologists frequently encounter turtles that are missing legs but have completely healed and recovered and seem well nourished and healthy. One potential problem is that female turtles missing hind legs may have difficulty nesting, although they may go through the instinctive motions of digging a nesting cavity, alternating between a good leg and the injured stump.

Injured turtles in the intermediate category have cracked shells in which the internal tissues may be visible through cracks or gaps where small shell pieces may be missing, but no major damage to internal organs or major blood loss has occurred. Turtles with broken shells have been repaired and gone on to live for many years, sometimes even laying eggs after they recover. A variety of splints can be devised to stabilize the shell, after cleaning the wound. Combinations of fiber tape, duct tape, epoxy, Superglue, and dental acrylic have all been used successfully to repair minor cracks.

Forest fires can be a threat to land-dwelling species such as box turtles if they cannot escape beneath the ground's surface. However, box turtles with heavily burned shells have been known to survive and later recover through shell regeneration.

Photo © Kurt Buhlmann

A turtle with a major injury, notably one smashed on a highway but still alive with its insides scattered on the road creates a more serious problem. No humane way to euthanize a seriously injured turtle on a highway has been agreed on. Some veterinarians or vet schools that have experience dealing with reptiles are willing to treat turtles that have been injured on the highway at no cost or, if appropriate, to euthanize turtles with major injuries beyond repair. Although an unpleasant task, the most effective way to put a severely injured turtle to death in a hopeless situation with no access to anesthetizing drugs is cervical dislocation, which, in the most straightforward language, means to cut off its head or sever the spinal cord.

Regrettably, a turtle's head can continue to function for several minutes after its removal, so this approach may not appear to be so humane. Unfortunately, many situations can arise in which even this action cannot be taken, and a badly injured turtle must be left to its own fate. People who have experienced such a dilemma are among the strongest advocates against building unnecessary convenience-only highways and are in favor of creating shoulder barriers and under-the-road passageways to prevent needless road kills of turtles and other wildlife on new and existing public road systems.

The immensity of the problem of injury and death of turtles on highways was brought dramatically to the attention of turtle biologists and the general public by Matt Aresco when he was a graduate student at Florida State University in 2000. By 2008, Aresco had rescued more than 8,800 turtles as they tried to cross "The World's Worst Turtle-killing Highway," heavily traveled U.S. Route 27 in Tallahassee, Florida, where it bisects Lake Jackson. Virtually all of these turtles rescued would have died if they had been allowed to cross the road. Aresco estimated that more than 325 turtles had lost their lives attempting to cross the busy highway during the same period despite efforts to prevent road kills by rescuing the turtles and by building road shoulder fences.

What should I do if I find a turtle crossing the road?

When a motorist sees a turtle crossing a highway, the inclination of many individuals is to move it off the road so that it will not be injured or killed by another vehicle. This is a perfectly acceptable act and makes a statement to others that someone is interested in the welfare of the animal. The first consideration for anyone observing a turtle crossing a highway is to be fully aware of their own safety and that of other motorists. It is obviously imperative that the vehicle be managed in a manner that does not distract or endanger the occupants or other drivers. Unfortunately, humanitarian attitudes toward turtles on roads must sometimes yield to situations in which saving the turtle is just too dangerous or risky under circumstances of fast or heavy traffic.

If the turtle can be approached safely and is uninjured, note the direction it was traveling when you first encountered it. Under most circumstances, the turtle had a chosen purpose or even a specific destination. If the turtle's purpose was to lay eggs, the female would be seeking higher ground and may not have a specific site selected but is searching for the proper location. In some instances, aquatic turtles, such as sliders (*T. scripta*) and painted turtles (*Chrysemys picta*), commonly migrate from one body of water to another and back for other reasons and know their destination as they travel. In either of these cases, releasing the turtle on the side of the highway in which it was headed would usually be the best choice. Where a turtle has reached an impasse, such as a high curb leading into a parking lot, it may be necessary to place the animal on the side from which it came. A common scenario with today's highway traffic is that both sides of the road are unsuitable for releasing a turtle because of urbanization. In such instances, it is sometimes better to pick the turtle up and release it farther down the highway in a more suitable habitat, but as close to where it was first seen as seems reasonable.

One cautionary note is that some turtles such as large snappers (*C. serpentina*) or softshells can bite or scratch and may be difficult to nudge to the side of the road. The simplest approach with a snapping turtle is to move quickly behind it, grab it at the base of the tail, and drag it to the road shoulder, always being careful to keep one's body parts clear from the aptly named snapping mouth. Clearly, such heroics should only be attempted by someone familiar with the behavior of these long-necked, pugnacious animals, which will bite and scratch to defend themselves and will not appreciate the efforts of a Good Samaritan.

What should I do if I find a turtle laying eggs?

Finding a turtle laying eggs is an exciting wildlife adventure and one that can be rewarding in several ways. If the turtle is not alarmed by your presence and continues to lay eggs, you may want to stand behind it and see whether you can count the number of eggs laid or simply watch the nest construction and departure of the female. Stay far enough away and move stealthily so that the turtle is not disturbed by your presence and appreciate this wonder of nature that has been working the same way for millions of years. One of the greatest threats to recently laid turtle eggs is that a predator will destroy the nest. Once the turtle has covered her nest and left the area, place a foot-square or larger piece of hardware cloth over the top of the nest and anchor it to discourage raccoons, foxes, and crows from digging up the nest. Predation on turtle nests usually occurs within the first day or so after the eggs are laid, so cover the nest as soon as possible. As time approaches for the eggs to hatch (usually more than 60 days), remove the protective hardware cloth so that the hatchlings can emerge from the soil and will not find themselves trapped.

What should I do if I find a baby turtle?

Finding a baby turtle can be a delightful experience. You are seeing native wildlife and learning something about the behavior and lifestyle of another animal. If you do not intend to keep the turtle for a pet or to show it to others, the simplest approach is to admire it briefly and then let it be on its way. A baby turtle could have recently left its nest on land and be headed toward the water or may have already made its way to the water's edge. If the turtle is clearly outside its native habitat, such as a turtle found crossing a highway or in a parking lot, the best rescue would be to place it in a wooded area or, if it is an aquatic species, at the edge of the closest suitable water. Baby turtles of many species make excellent pets, but the owner must be aware that it will not stay small and that some species are protected and permits are required to hold them in captivity.

How can I see turtles in the wild?

In most areas of the world where turtles occur, the most dependable opportunities for seeing them in the wild are around large rivers or lakes where turtles bask on logs or rocks. Using a pair of binoculars can en-

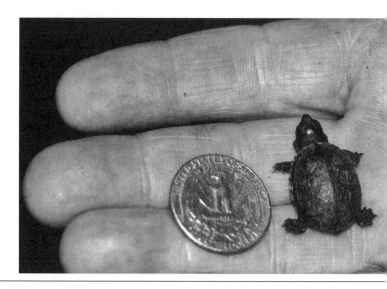

Eastern mud turtles lay their eggs in the spring and early summer. The tiny babies hatch two to three months later and then spend the entire winter in underground nests; they emerge the following spring.

SREL file photo

hance the experience, as those basking nearby are likely to retreat into the water if they see someone approaching. Bridges that span large rivers are particularly good vantage points when river turtles are present, particularly if there are good basking sites, such as rocky outcrops present. If the bridge permits pedestrian traffic so that a person can stand in the center and observe the shorelines, a careful observer may be able to see turtles on logs, brush, and other vegetation along the edges. Many nature centers and outdoor camps have wildlife field trips, and turtles often are among those animals that can be observed. In coastal areas, especially in South Carolina, Georgia, and Florida, during sea turtle nesting season, you may encounter a female nesting as you walk along an unlighted beach. Later in the season, if the timing is right, a beachgoer might be fortunate enough to see dozens of hatchlings as they emerge from an earlier laid nest and make their way to the water in the moonlight. Another place where a variety of freshwater turtle species can be seen in their natural habitat is in the clear springs in northern Florida. Many of these springs allow rafting, snorkeling, and underwater diving, whereupon a quiet observer can view turtles feeding and swimming in the clear water.

Should people feed turtles in lakes?

Although some wildlife purists might disagree, feeding turtles in lakes is generally not a problem for the turtles or other wildlife as long as it is not overdone. Turtles can go for long periods without eating, and therefore, such food would only be a supplement to their natural diet and would not result in a dependency on being fed in a natural lake. Turtles are not

In many areas of the world where turtles occur, a dependable opportunity for seeing them in the wild is around large rivers or lakes where they bask on logs or rocks. Photo © David Scott

like alligators that can become a threat to humans when they are fed on a regular basis, and no turtle is likely to suffer from a health standpoint because of supplemental feeding as is sometimes true with wild mammals that are overfed. In most situations, feeding turtles around a lake provides an opportunity to observe them and appreciate them without negative consequences to the animals themselves.

Chapter 9

Turtle Problems (from a human viewpoint)

Are turtles pests?

No. Native turtles living in natural environments should never be considered pests of any sort. Turtles are not dangerous to humans, not even to children, under ordinary circumstances. When someone gets bitten while pestering a snapping turtle (*Chelydra serpentina*), it does not mean the turtle is the pest. Animals defined as pests have behaviors that are annoying or harmful to humans in some way. Some of the obvious pesky behaviors that turtles seldom if ever engage in are entering people's houses uninvited, raiding garbage cans, biting people unprovoked, or causing allergic reactions. Turtles are not like some birds and mammals that serve as vectors to transmit diseases to humans. Tortoises and even box turtles (*Terrapene*) may raid your garden or flower bed and eat vegetables or fruit, such as lettuce, tomatoes, and fallen apples or pears, or ornamental shrubs, but their effect can usually be readily controlled by limiting their access. Turtles also do not make annoying noises or, except for rare instances with giant tortoises, leave unpleasant signs of their former presence.

Do turtles reduce the number of fish in lakes and rivers?

No scientific evidence has been presented to document that any kind of turtle appreciably reduces the population size of game fish in a lake or river. This is not to imply that turtles do not catch live fish in some situations, using a variety of techniques, which include tongue luring by alligator snappers (*Macrochelys temminckii*), sit-and-wait ambush attacks by

The impressively long neck of snake-necked turtles is presumably used to search for prey beneath difficult-to-reach crevices underwater or to strike rapidly to capture aquatic prey.

Photo © Cris Hagen

matamatas (*Chelus fimbriata*) and softshell turtles (genus *Trionyx*), and rapid strikes by long-necked species, such as chicken turtles (*Deirochelys reticularia*) of North America and some of the snake-necked turtles of South America (genus *Hydromedusa*) and Australia (genus *Chelodina*).

However, turtles eat fish that are probably already dead from causes other than being caught by a turtle; in fact, some lakes might look and smell bad were it not for the turtles acting as scavengers to remove dead fish. Unfortunately, turtles become scapegoats for unsuccessful fishermen, who blame turtles when the fish are not biting.

Do turtles kill ducks in ponds?

Yes, large individuals of some turtle species may eat a young duckling or even a full-grown duck swimming on the surface, as they would any other prey animal. The primary culprits of such behavior, which is entirely natural and nothing to be alarmed about (unless you are the unlucky duck), are common snapping turtles (*C. serpentina*). Most other turtles do not get large enough to capture baby ducks and geese. To a snapping turtle underwater, any small animal on the surface or underwater, whether it's a snake, a baby alligator, a duck, a muskrat, or even a smaller turtle, is viewed as available prey. Rarely would the density of turtles, however, be great enough to result in long-term negative effects on the local duck population.

Do turtles have diseases and are they contagious?

Turtles are noted for having a variety of afflictions that are characteristic of certain species or regions. Among the most notorious is *Salmonella*. Turtles are considered by some health agencies to be a potential source of *Salmonella* poisoning that may not have noticeable effects on the turtle as a carrier but that can be transmitted to humans. *Salmonella* bacteria cause an unpleasant digestive system ailment (salmonellosis) characterized by diarrhea, fever, and cramps within a day of exposure. The likelihood of developing symptoms after more than a day or two is unlikely to nonexistent. Other kinds of reptiles besides turtles carry *Salmonella*, but healthy reptiles are usually not themselves affected.

Many *Salmonella* strains exist, and the virulence to humans varies greatly. A person's health, age, and physiological condition can determine one's response to exposure. Infants, the elderly, and people with chronic illnesses or immune deficiencies may be more susceptible to *Salmonella* than healthy adults, and the consequences can be more severe; nonetheless, the risk of contracting salmonellosis is still low. Some authorities claim that children younger than eight years old should avoid contact with turtles that could carry *Salmonella*. Others have more lenient opinions, suggesting supervised handling of pet turtles by children and a requirement that hands be kept away from the mouth, nose, eyes, and food until they can be washed with soap and water. Salmonellosis can be serious or even fatal in people with compromised immune systems or other forms of poor health, but few people require medication to recover safely. Indeed, mild cases may not even be recognized, manifesting themselves as only a mild digestive upset.

The possibility of contracting *Salmonella* from handling wild or pet turtles exists, but the risks are so low that worrying about the possibility causes more stress than is warranted. A noteworthy observation is that staff and students of the University of Georgia's Savannah River Ecology Laboratory in South Carolina, where literally thousands of turtles of dozens of species have been studied and handled since the late 1960s, have never experienced a confirmed case of salmonellosis.

Most turtle biologists believe that the risk of salmonellosis in most instances is too small to warrant concern about keeping or picking up turtles. If you like to handle turtles, or want your child to have one for a pet, be aware of any hazards, but weigh the cost-benefit odds. The benefits will usually win. Parents should educate their children about potential hazards and precautions, such as always washing their hands after handling a turtle, but then be allowed to make their own choices about which risks to take for themselves or their children. Understanding the importance of cleanliness to reduce the threat of *Salmonella* poisoning or other health hazards is critical for any pet turtle owner and is common sense for one's health in general. Thus, proper care of turtles in captivity and proper hygiene after handling them can minimize and potentially eliminate the possibility of contracting salmonellosis.

Turtles can contract a variety of other notable diseases that are not

Although generally not a major threat or hazard to turtles, ocean-going turtles and diamondback terrapins that live in brackish water often acquire barnacles that grow on their shells. Photo © Cris Hagen

known to be contagious to humans but that can be transmitted among individuals in turtle populations. Upper respiratory tract disease, known as URTD among turtle biologists, is a highly contagious, pneumonia-like infection of respiratory tissues in desert tortoises (*Gopherus agassizii*) and gopher tortoises (*Gopherus polyphemus*). A species of bacteria (*Mycoplasma agassizii*) has been identified as the cause of the disease.

Fibropapilloma, a viral disease, affects sea turtles and causes unsightly tumors to appear on the neck, eyes, and other body parts of green sea turtles (*Chelonia mydas*). The disease was first noted as prevalent in green sea turtles in Hawaii in the late 1900s and later in populations in Florida. Individual turtles with fibropapilloma can recover, but whether fibropapilloma will become a widespread problem for green sea turtles worldwide, debilitate the entire species, or develop as a problem in other species of turtles is yet to be determined.

Is it safe to eat turtles?

Records exist of people becoming ill after eating certain turtles. Although the small, heavy-shelled common box turtle (*Terrapene carolina*) is not one most people would ever consider eating in the first place, box piece of butter, and warm milk enough to make it quite soft. Put it into a buttered dish, rub butter over the top, shake over a little sifted flour, and bake about thirty minutes, and until a rich brown. Make a sauce of drawn butter, with two hard-boiled eggs sliced, served in a gravy-boat.

CODFISH STEAK. (New England Style.)

Select a medium-sized fresh codfish, cut it in steaks cross-wise of the fish, about an inch and a half thick; sprinkle a little salt over them, and let them stand two hours. Cut into dice a pound of salt fat pork, fry out all the fat from them and remove the crisp bits of pork; put the codfish steaks in a pan of corn meal, dredge them with it, and when the pork fat is smoking hot, fry the steaks in it to a dark-brown color on both sides. Squeeze over them a little lemon juice, add a dash of freshly ground pepper, and serve with hot, old-fashioned, well-buttered Johnny Cake.

SALMON CROQUETTES.

One pound of cooked salmon (about one and a half pints when chopped), one cup of cream, two tablespoonfuls of butter, one tablespoonful of flour, three eggs, one pint of crumbs, pepper and salt; chop the salmon fine, mix the flour and butter together, let the cream come to a boil, and stir in the flour and butter, salmon and seasoning; boil one minute; stir in one well-beaten egg, and remove from the fire; when cold make into croquettes; dip in beaten egg, roll in crumbs and fry. Canned salmon can be used.

Shell=Fish.

STEWED WATER TURTLES, OR TERRAPINS.

Select the largest, thickest and fattest, the females being the best; they should be alive when brought from market. Wash and put them alive into boiling water, add a little salt, and boil them until thoroughly done, or from ten to fifteen minutes, after which take off the shell, extract the meat, and remove carefully the sand-bag and gall; also all the entrails; they are unfit to eat, and are no longer used in cooking terrapins for the best tables. Cut the meat into pieces, and put it into a stew-pan with its eggs, and sufficient fresh butter to stew it well. Let it stew till quite hot throughout, keeping the pan carefully

covered, that none of the flavor may escape, but shake it over the fire while stewing. In another pan make a sauce of beaten yolk of egg, highly flavored with Madeira or sherry, and powdered nutmeg and mace, a gill of currant jelly, a pinch of cayenne pepper, and salt to taste, enriched with a large lump of fresh butter. Stir this sauce well over the fire, and when it has almost come to a boil, take it off. Send the terrapins to the table hot in a covered dish, and the sauce separately in a sauce-tureen, to be used by those who like it, and omitted by those who prefer the genuine flavor of the terrapins when simply stewed with butter. This is now the usual mode of dressing terrapins in Maryland, Virginia, and many other parts of the South, and will be found superior to any other. If there are no eggs in the terrapin, "egg balls" may be substituted. (See recipe).

STEWED TERRAPIN, WITH CREAM.

Place in a sauce-pan, two heaping tablespoonfuls of butter and one of dry flour; stir it over the fire until it bubbles; then gradually stir in a pint of cream, a teaspoonful of salt, a quarter of a teaspoonful of white pepper, the same of grated nutmeg, and a very small pinch of cayenne. Next, put in a pint of terrapin meat and stir all until it is scalding hot. Move the sauce-pan to the back part of the stove or range, where the contents will keep hot but not boil; then stir in four well-beaten yolks of eggs; do not allow the terrapin to boil after adding the eggs, but pour it immediately into a tureen containing a gill of good Madeira and a tablespoonful of lemon juice. Serve hot.

STEWED TERRAPIN.

Plunge the terrapins alive into boiling water, and let them remain until the sides and lower shell begin to crack—this will take less than an hour; then remove them and let them get cold; take off the shell and outer skin, being careful to save all the blood possible in opening them. If there are eggs in them put them aside in a dish; take all the inside out, and be very careful not to break the gall, which must be immediately removed or it will make the rest bitter. It lies within the liver. Then cut up the liver and all the rest of the terrapin into small pieces, adding the blood and juice that have flowed out in cutting up; add half a pint of water; sprinkle a little flour over them as you place them in the stew-pan; let them stew slowly ten minutes, adding salt, black and cayenne pepper, and a very small blade of mace; then add a gill of the best brandy and half a pint of the very best sherry wine; let it simmer over a slow fire very gently. About ten minutes or so, before you are ready to dish them, add half a pint of rich cream, and half a pound of sweet butter, with flour, to prevent boil-

ing; two or three minutes before taking them off the fire, peel the eggs carefully and throw them in whole. If there should be no eggs use the yolk of hens' eggs, hard boiled. This receipt is for four terrapins.

—*Rennert's Hotel, Baltimore.*

BOILED LOBSTER.

Put a handful of salt into a large kettle or pot of boiling water. When the water boils very hard, put in the lobster, having first brushed it, and tied the claws together with a bit of twine. Keep it boiling from 20 minutes to half an hour in proportion to its size. If boiled too long the meat will be hard and stringy. When it is done take it out, lay it on its claws to drain, and then wipe it dry.

It is scarcely necessary to mention that the head of a lobster, and what are called the lady-fingers, are not to be eaten.

Very large lobsters are not the best, the meat being coarse and tough. The male is best for boiling; the flesh is firmer, and the shell a brighter red; it may readily be distinguished from the female; the tail is narrower, and the two uppermost fins within the tail are stiff and hard. Those of the hen lobster are not so, and the tail is broader.

Hen lobsters are preferred for sauce or salad, on account of their coral. The head and small claws are never used.

They should be alive and freshly caught when put into the boiling kettle. After being cooked and cooled, split open the body and tail, and crack the claws, to extract the meat. The sand pouch found near the throat should be removed. Care should be exercised that none of the feathery, tough, gill-like particles found under the body shell get mixed with the meat, as they are indigestible, and have caused much trouble. They are supposed to be the cause of so-called poisoning from eating lobster.

Serve on a platter. Lettuce, and other concomitants of a salad, should also be placed on the table or platter.

SCALLOPED LOBSTER.

Butter a deep dish, and cover the bottom with fine bread-crumbs; put on this a layer of chopped lobster, with pepper and salt; so on alternately until the dish is filled, having crumbs on top. Put on bits of butter, moisten with milk, and bake about twenty minutes.

DEVILED LOBSTER.

Take out all the meat from a boiled lobster, reserving the coral; season highly with mustard, cayenne, salt and some kind of table sauce; stew until well mixed,

Among the earliest published recipes for turtles were those in the shellfish section of *The White House Cookbook* of 1887 by Mrs. F. L. Gillette. The callous recommendation of dropping the turtle alive into boiling water indicates the cultural attitudes of the times. Photos © Margaret Wead

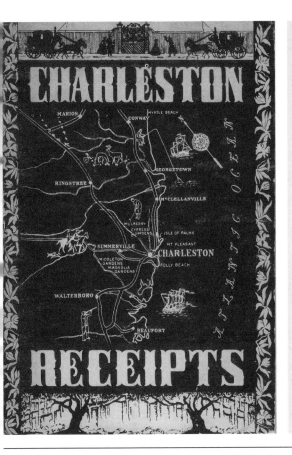

CHARLESTON RECEIPTS

Cooter Soup

1 large or 2 small	Red pepper to taste
"yellow belly" cooters	3 tablespoons dry sherry
(preferably female)	4 quarts of water
1 large onion, chopped	1 small Irish potato, diced
Salt to taste	12 whole cloves
2 teaspoons allspice	2 tablespoons Worcestershire
	Flour to thicken

Kill cooter by chopping off head. Let it stand inverted until thoroughly drained, then plunge into boiling water for five minutes. Crack the shell all around very carefully, so as not to cut the eggs which are lodged near surface. The edible parts are the front and hind quarters and a strip of white meat adhering to the back of the shell, the liver and the eggs. Remove all outer skin which peels very easily if water is hot enough. Wash thoroughly and allow to stand in cold water a short while, or place in refrigerator over night.

Boil cooter meat, onion and potato in the water, and cook until meat drops from bones—about 2 hours. Remove all bones and skin and cut meat up with scissors. Return meat to stock, add spices and simmer. Brown flour in skillet, mix with 1 cup of stock to smooth paste and thicken soup. Twenty minutes before serving add cooter eggs. Add sherry and garnish with thin slices of lemon. Serves 6-8.

Mrs. CLARENCE STEINHART (Kitty Ford)

The recipe for yellow-bellied slider turtle in *Charleston Receipts*, published in the 1950s by the Junior League of Charleston, South Carolina, recommended a more humane approach for preparing the turtle for a meal. Photos © Margaret Wead

turtles sometimes feed on poisonous mushrooms without being adversely affected. The toxic compounds in the mushroom are still present in the turtle and can concentrate in the muscle tissue. Therefore, a person eating box turtle meat soon after the animal has consumed a poisonous mushroom could have adverse reactions. Other turtles are also known to tolerate eating items that contain material poisonous to humans. If a toxic material is sequestered in the muscle tissue of a turtle, another animal eating the turtle could be affected, unless they too are immune to the poison. According to several reputable reports, people have become sick or even died after eating hawksbill sea turtles (*Eretmochelys imbricata*), presumably because the turtle had eaten sponges toxic to people but not to the turtle.

For most species of turtles large enough to be prepared and cooked (for example, adult snapping turtles [*C. serpentina*], softshell turtles, diamond-back terrapins [*Malaclemys terrapin*], and gopher tortoises [*Gopherus polyphemus*]), there is no inherent reason the meat on the legs and attached to the shell would not be suitable for consumption. The famous 1950 cookbook

This "wild caught turtle meat" (species unknown, but most likely common snapping turtle) was for sale in the gourmet meat section at a national supermarket chain in South Carolina in March 2009. Turtle meat is considered an exotic food item and is regulated by the U.S. Food and Drug Administration rather than the U.S. Department of Agriculture. All turtle meat is included (oddly enough) in the seafood category (FDA, Title 21, *Code of Federal Regulations*, Pt. 123), perhaps because green sea turtle meat was widely eaten as a delicacy before sea turtle harvesting was banned. All seven species of sea turtles are included on the IUCN endangered species list. Photo © Hayley McLeod

Charleston Receipts had a receipt (recipe) in the fish section for Bluff Plantation cooter pie that included the liver and eggs as well as the meat. *The White House Cookbook* (1887), by Mrs. F. L. Gillette, featured "Stewed Water Turtles, or Terrapins" in the "Shell-Fish" section. Both recipes called for adding adequate alcoholic flavoring, and the latter was unwavering in instructions that the turtle should be dropped "alive" into "boiling water." We do not recommend this recipe.

An unfortunate consequence of the worldwide environmental pollution of aquatic habitats, beginning in the twentieth century and continuing into the twenty-first century, is that many aquatic animals, including ocean tuna and freshwater fishes in many rivers and large lakes, are no longer considered safe to eat. Turtles living in aquatic areas with industrial pollution of toxic chemicals, heavy metals, and certain organic and inorganic compounds, can survive, but the contaminants can be sequestered in their muscle tissue, much like the mercury toxicity resulting from eating some kinds of fish. Because turtles live for such a long time, those from contaminated habitats can accumulate and harbor toxic materials over many years, and these can be passed on directly to humans who eat the animals.

Turtles: The Animal Answer Guide

What should I do if I get bitten by a turtle?

The first step in dealing with a turtle bite often requires little more than common sense—get the turtle to let go. The old adage that a turtle will hold on until it thunders is of course absurd but a painful bite by even a small turtle that simply will not let go sometimes makes one wonder whether it might take that long. Under most circumstances, anyone being bitten by a turtle has "bothered" the turtle or intentionally put themselves within striking distance. A turtle bites because it feels threatened and has decided to fight to defend itself. Therefore, in most instances, if an aquatic turtle is placed in water such as at the edge of a lake or stream, it will let go once it realizes it can escape. However, because most turtle bites occur in situations in which the turtle is being handled or provoked, a pond or lake may not be handy. In such an instance, placing a small turtle in a sink filled with water may be an effective alternative that results in the animal releasing its grip.

In one documented case with a large snapping turtle (*C. serpentina*) that was two feet underwater in the mud on the river bottom, someone was bitten on the finger and the turtle would not release its grip. The person knew that the object on the bottom of the river was a turtle and had attempted to catch it but unfortunately had reached for the wrong end of what turned out to be a large snapping turtle. After several agonizing minutes in which the turtle maintained its painful grip, the turtle biologist finally determined that he would have to take the animal ashore to get help in extracting his fingers from the turtle's mouth. Fortunately, when the turtle was lifted above the surface of the water, having maintained a strong grip up to that point, it opened its mouth giving the chagrined biologist an opportunity to remove his lacerated, but unbroken, fingers. The turtle was taken to shore, marked for future identification, and then released back in the river. Lacerations on the fingers were extensive and the finger bone was actually visible, although the turtle's bite did not break the bone. This, of course, calls into question the stories that large common snappers (*C. serpentina*) are able to break a broom handle in half, as a standard wooden broom handle is much thicker and stronger than a human finger.

Most other situations in which people are bitten by turtles, usually small ones, result in pinching or minor bruising of whatever body part was unlucky enough to have gotten within range. The results are usually not serious but may result in small cuts from the sharp beak edges and can be treated with soap and water and an antiseptic. Giant tortoises being hand-fed fruit or vegetables have been known to accidentally bite the person's hand that feeds them. Although such turtle bites are not intentionally malicious or aggressive, they can be painful and actually break the skin. Infection or other complications from a turtle bite that has been properly cleaned should be no more of a problem than any other cut or laceration and would not require special treatment, unlike more potentially negative wildlife encounters such as venomous snake bite; stings of a scorpion, bee, or wasp; or the possibility of rabies from some mammal bites.

Chapter 10

Human Problems (from a turtle's viewpoint)

Are any turtles endangered?

The formal classification of animals as endangered involves processes that are based on the biology and conservation status of the species themselves but are also influenced by politics that results in some species being listed when they should not be and others not being listed when they should. An additional factor is that even experts do not always agree on how data should be interpreted and what the status of a species should be. However, formal listing and categorization of species in terms of their vulnerability is done on an international scale by the International Union for Conservation of Nature (IUCN). The United States has its own independent endangered species list that includes several turtles.

The IUCN recognizes several levels of peril for species, and turtles have been identified in several categories that include, in order from bad to worst, vulnerable (VU), endangered (EN), critically endangered (CR), and extinct in the wild (EW). An additional category known as "data deficient" (DD) means that no suitable information is available to declare the conservation status of the species. The IUCN Red List is understandably dynamic because it is constantly changing based on new information for some species and the actual change in status of others. The 2007 Red List included turtle species in the categories of vulnerable (57), endangered (46), critically endangered (24), and extinct in the wild (1), in addition to those in the data deficient category.

The U.S. Fish and Wildlife Service, in accordance with the Endangered Species Act, listed 13 species of turtles in 2008. The most serious categories that apply to the turtles based on their conservation status include en-

A threat that supersedes even the loss of habitat is the removal of turtles from the wild for the food market at unsustainable levels.

Photo © Rebecca Yeomans

dangered (5 species) and threatened (8 species), of which 6 are sea turtles, 2 are tortoises, 4 are river species, and 1 is the bog turtle, which is threatened in part of its range. As with the IUCN list, changes are proposed every year, and so the species composition can change, although not all proposals are accepted.

Whether or not particular species are listed on an official registry, many species of turtles are unquestionably in grave danger of declining numbers and many are threatened with extinction if harmful conditions and practices continue unabated. The major threats are alteration and destruction of their nesting areas, food sources, and basic habitats. An additional threat that supersedes even the loss of habitat is the removal of turtles from the wild for the food market at unsustainable levels. Kurt Buhlmann, at that time on the staff of Conservation International, gave a particularly sobering account about the status of freshwater turtles in Asia when he visited Cambodia in 2000.

> I attended a workshop in Phnom Penh, Cambodia, with forty people from sixteen countries to discuss international trade and exploitation of freshwater turtles and tortoises in Asia. Most Americans view turtles as interesting and

harmless creatures; I learned that to many Asians they represent food and medicine.

As I wandered through the markets of Phnom Penh, incredibly crowded and bustling places, I met a wildlife trader with eighteen rice sacks, each containing more than a hundred pounds of live turtles. They had been collected throughout the countryside by local people and brought to the trader for cash. He was transporting them to the Vietnamese border where they would be taken by someone else to China.

In China, with more than a billion citizens, people pay handsomely for the opportunity to eat a turtle or use the ground up shells in traditional medicine concoctions. Here I learned the truly frightening reality: this scenario is taking place daily in many cities not only in Cambodia but also in Vietnam, Malaysia, Indonesia, Bangladesh, Thailand, and elsewhere.

On the island of Sumatra, one observer with TRAFFIC, an organization that monitors international trade in wildlife, documented 25 tons of live turtles per week leaving one port by ship. The eventual destination of the turtle shipment: food markets of China. Poor villagers have become turtle collectors because, depending on the species, one turtle may provide the equivalent of their usual income for a day, a week, even a year.

Clearly, such economic incentives will lead to major negative effects on wild populations of turtles. Wholesale removal of turtles from their natural habitats is an unsustainable process and, for most species, can only lead to extirpation locally and, for some, ultimately extinction. The Asian turtle crisis can only be stopped by government intervention and shifts in cultural attitudes. Without any action, disappearance of the turtles from the wild is inexorable.

Will turtles be affected by global warming?

Turtles are an excellent example, for several reasons, of how certain wildlife could be affected by a measurable rise in the earth's temperature. Despite the politically charged atmosphere surrounding attitudes about whether global warming or, as some prefer, climate change, is a problem or whether industrial nations are the primary cause, many species of turtles could be affected by a rise in surface temperatures worldwide.

Climate changes associated with global warming could include widespread shifts in seasonal patterns of temperature and rainfall, long-term droughts, and rises in water and soil temperatures during critical life history phases of some species. Each of these climatic changes could affect the ecology, behavior, and physiology of some species of turtles. During droughts, some species of turtles, such as slider turtles (*Trachemys scripta*), move overland to other bodies of water as the aquatic site they live in dries. If regional droughts are extended because of global warming, turtle populations in habitable sites would remain at higher than normal densities, resulting in the potential for limited resources, including food and nesting sites, and a concentration of individuals for certain forms of predation.

An increase in predation could result from more frequent exposure of individual turtles during overland movement. Long-term droughts could also result in extended hibernation/aestivation that could affect survivorship patterns. In addition to changes that could occur among turtles and other wildlife species in the reduction of resources, hibernation/aestivation behavior, and movement patterns, turtles could be affected in another way. A rise in soil temperature could affect the sex ratio of hatchling turtles.

Unlike mammals and birds, most turtles do not have sex chromosomes that predetermine the sex of the individual. Although environmental variables rather than genetics (environmental sex determination [ESD]) may be involved in determining the sex of a baby turtle when it emerges from the egg, temperature during incubation appears to be the overriding factor. Among most species of turtles that have been studied closely, the temperature at which the eggs are incubated affects whether the hatchlings emerging from a clutch of eggs are males or females.

The phenomenon called temperature-dependent sex determination (TSD) occurs in turtles during the middle one-third of incubation and reflects the average temperature during that period. In simplest terms, when turtle eggs are raised at warm temperatures they produce female babies. Cooler temperatures typically produce males. However, in some species, extreme high or low temperatures produce one sex and intermediate temperatures produce the other. Thus, much variation exists among species, and not all have been carefully determined with incubation experiments. Nonetheless, if environmental temperatures were raised significantly so that incubation temperatures during nesting also increased, only one sex of some species would be produced. The trajectory presumably would lead to extinction.

Are turtles affected by pollution?

Although how turtles are directly affected by the uptake of contaminants into their bodies is still under study by biologists, turtles living in polluted waters could potentially be a problem for humans who eat them. Dangerously high levels of mercury have been found sequestered in the muscle tissue of turtles living in rivers polluted by industries in Asia. Likewise, the accumulation of organic compounds known as PCBs (polychlorinated biphenyls) in diamondback terrapins (*Malaclemys terrapin*) from an

Like other wildlife, turtles are susceptible to aquatic pollution. More important, from a human consumption standpoint, turtles often survive in contaminated areas although they may have concentrated levels of heavy metals or other toxic materials in their muscle tissues. SREL file photo

area in Brunswick County, Georgia, has also been identified. The Asian species and the diamondback terrapins that harbor these contaminants are species eaten by people in certain regions and subsequently could result in the consumption of hazardous levels of these contaminants.

Radioactively contaminated slider turtles (*T. scripta*) with extraordinarily high levels of radioactive cesium-137 and strontium-90 were observed over a one-year period by David Scott at the Savannah River Ecology Laboratory, where the turtles were kept in an isolated experimental pond. During the time they were sequestered, the turtles lost approximately half of their body burden of cesium every two months and half of their strontium in a year, indicating that the biological half-life of some contaminants in turtles is relatively short, a possible explanation for the lack of effects on the observed animals.

Turtles may be affected by pollution in other ways besides uptake into the body and by different kinds of contaminants in the environment. Ironically, for painted turtles (*Chrysemys picta*) in the Kalamazoo River in Michigan, an argument could be made that river pollution worked to the turtles' advantage. The effluents from paper mills and other industries along the river were unregulated at the time and clearly were responsible for the elimination of many of the fish and wildlife species that naturally occurred there. However, painted turtles were present in some stretches of the river where no fish were able to live, and yet the turtles appeared to thrive on the larvae of midge flies, known as chironomid worms, that were able to live in the contaminated systems. The turtles in the Kalamazoo River grew more rapidly, reached larger sizes, and laid more eggs than painted turtles in nearby natural areas in Michigan.

Another effect of contamination on turtles of aquatic systems was suspected genetic damage to turtles living in a reservoir on the Savannah River Site in South Carolina with low-level radioactive contaminants from a nuclear production reactor. An examination of chromosomes of slider turtles living in the lake where the sediments were contaminated by a radioactive spill of cesium 30 years earlier gave evidence of chromosome breakage and genetic reconfiguration. No other noticeable effects of developing embryos, survivorship of juveniles or adults, or egg-laying productivity were documented. Also, genetic repair is known to occur in animals affected in the manner observed in the turtles, so permanent chromosomal damage may not have occurred in the turtles in the population. The genetic changes that occurred in somatic cells of the body were not transmitted to offspring. No obvious differences in individual appearance or population structure and demography were apparent between turtles in the reservoir and populations of slider turtles in other aquatic habitats in the surrounding regions.

Extensive evidence from scientific studies exists that demonstrates that turtles living in polluted aquatic habitats accumulate heavy metals, radioactive isotopes, and other contaminants present in the system. However, documentation that turtles are severely affected or impaired in any way by the presence of contaminants in their bodies is still forthcoming. The shell of turtles may be a reason for this group of animals to be minimally affected by certain environmental pollutants because they may be able to rid themselves of heavy metals, radioactive materials, or other contaminants. Research is still under way to confirm whether some toxic materials can be transferred from other parts of the body to the shell, where they are ultimately shed in those species of turtles, such as slider turtles (*T. scripta*) and snappers (*Chelydra serpentina*), that shed their scutes.

Why do people hunt and eat turtles?

People hunt turtles and collect their eggs for food around the world, purely for sustenance in some societies, as delicacies in some cultures, and as perceived medicines or aphrodisiacs in others. Eating turtle eggs and larger individual turtles was a widespread practice among primitive societies living on tropical oceanic islands and around large rivers and lakes. Turtles were considered an edible component of the available fauna and were consumed at a sustainable rate in situations where humans lived off the land and did not exploit turtles as a commodity. However, as human populations have exploded, slowly maturing animals such as turtles cannot

Turtles: The Animal Answer Guide

keep up with the demand on their numbers for use as food. Hence, they are experiencing declines throughout the world.

Are "tortoiseshell" items actually made from turtle shells?

Authentic tortoiseshell, also called *carey*, is made from the shell of hawksbill sea turtles (*Eretmochelys imbricata*), which are on the U.S. federal list of endangered and threatened species. Authentic horn-rimmed glasses, tortoiseshell combs, and other artifacts were formerly made from the decorative scutes of these animals. In the United States and most other countries, buying or selling items made from tortoiseshell is now illegal. Much that appears to be tortoiseshell items nowadays is in fact plastic.

Why do so many turtles get hit by cars?

Because of the extensive highway systems in North America and the propensity of both aquatic and terrestrial turtles to move between habitats, tens of thousands of turtles are killed every year on U.S. highways. Many of these are recent hatchlings that have left the nest and are headed toward the closest wetland. If a road has been built between the wetland and the upland nesting site, numerous fatalities can be expected. Also females are killed on the way to the nesting habitat, which means that their eggs are also destroyed. Another reason turtles are frequently found on roads is that males of some species travel overland extensively between wetlands in search of females during the mating season, which is usually in the spring. A study by Steve Morreale and others at SREL documented that male slider turtles (*T. scripta*) moved longer distances than females in most months, including between bodies of water, which could result in their crossing highways and being hit by cars.

Female turtles of many species perceive road shoulders, even along busy highways, to be suitable nesting sites. Road shoulders are often higher than surrounding wetland areas and have soft soil. Many times a turtle will move from one side of the highway to reach the other shoulder in search of the ideal nesting spot, thereby exposing itself to oncoming traffic. James Gibbs and David Steen of State University of New York examined records for 16 species of turtles native to the eastern United States and concluded that a greater number of females were killed each year compared with males.

An additional component that explains why so many turtles are hit by cars is that adult turtles may have a false sense of security about predators because of their thick protective shell. Unfortunately, most turtles that cross highways do not reach sizes large enough to keep cars and trucks from crushing them. Notable exceptions are certain large species such as the giant tortoises on the roads of Santa Cruz in the Galápagos that reach sizes at which an aware motorist would avoid them. Under natural conditions throughout the evolutionary history of turtles, until modern times, a turtle crossing an open space was able to ward off most potential predators by the protective features of a suit of armor that is ineffective against an automobile tire.

Are boats dangerous for aquatic turtles?

The propellers of outboard motors can cause damage or death to aquatic turtles, and the phenomenon is probably much more common than recognized because of the difficulty of obtaining data. When approached by a boat, a turtle's normal response is to dive below the surface and head to the bottom in most instances. However, a fast-moving motorboat can overtake a diving turtle before it reaches a safe depth so that the propeller catches the animal in midwater. In one study along the South Carolina coast, measurements were made of shell damage to diamondback terrapins (*M. terrapin*) in a river with moderate boat traffic. The study showed that terrapins were missing appendages and had broken shells that had healed. Most of the leg injuries and missing legs were apparently due to aquatic predators, such as sharks or large blue crabs, whereas the shell damage was determined to be from boats. Equal damage could be expected to occur in rivers and lakes where turtles encounter boat traffic, but studies to make the determination have not been done. Although documentation of mortality by turtles from boat props is not available, death from injuries undoubtedly occurs in some situations. The incidence of recorded injuries represents only a small percentage of the turtles that are actually hit by outboard motors.

How are turtles affected by litter?

Litter generally indicates the lack of respect people have for their environment. Litter along highways, waterways, and natural areas is unsightly and represents someone's inappropriate behavior. However, litter can also be more than simply an assault on environmental cosmetics. Some turtles can actually die because of litter: the primary culprit is plastics, which turtles, accustomed to eating floating objects, may consume. There are documented cases of leatherback sea turtles (*Dermochelys coriacea*) that have eaten plastic bags floating in the ocean that they presumably mistook for jellyfish, a common food item of the species. Ironically, the heaviest leatherback sea turtle ever reported was found dead in Wales. Dissection showed that plastic had lodged in its digestive tract, which may have prevented it from eating, leading to its death. Plastic soft drink and beer rings also cause problems for turtles. If a small turtle crawls through a ring, the material can be lodged around the body of the turtle and remain there even though the turtle continues to grow throughout its lifetime, with the result that its body becomes constricted in the area where the plastic ring is pinching it. Such damage is permanent once the shell is hardened, often resulting in a badly deformed turtle.

What can an ordinary citizen do to help turtles?

When people become aware that turtles or any other group are endangered, they are usually concerned, and many want to know what they can do to help. The problem would be solved if everyone had the attitude that all turtles were special and that humans have a responsibility to protect them at all costs. Unfortunately, we will never have a world in which all people think all turtles should be protected from actions and activities that eliminate individuals or even entire populations or species of turtles. However, the more people there are who feel that turtles are a part of our natural world and are worth keeping around forever, the greater the sense of stewardship will be in a society. The attitude of good stewardship in a majority of humans in a region, a state, or a country can lead to protective measures. Some people who do not have an innate or learned attitude that humans have a responsibility to protect all species of wildlife will respond to legislation, if not education.

Rules and regulations that protect turtles have been implemented in many U.S. states and on an international scale for some countries. Such rules can only be made if people in a society are vocal about passing them and convince the appropriate politicians to support them. Often, rules to protect turtles are made but the enforcement, which all too often is inadequately funded, does not follow. Turtles are protected on paper but not in the wild. People with a strong interest in preserving and protecting turtles must continue to shift public attitudes toward one of responsible stewardship so that improper actions against turtles and other wildlife are viewed with public scorn, regardless of whether a law enforcement official is present.

Probably the best effort a single individual can make in protecting turtles is to change the attitudes of other individuals. One approach is to teach others, particularly children, that turtles are part of the native fauna in most areas and deserve to exist without threats to their lives or habitats and that such an effort is the right thing to do. The removal of a single turtle for a child to keep as a pet will do no harm to a population of turtles and may lead to the child having the necessary awareness of turtles and appreciation for their existence. Consequently, working toward legislation that protects populations and species of turtles and their required habitats is often more important than the fate of a single individual turtle. For example, many of us will always remove a turtle from a road, if we can do so safely and without endangering ourselves, but anyone who could prevent a highway from being built in the first place would do far more for any turtle populations in the area.

Chapter 11

Turtles in Stories and Literature

What roles do turtles play in religion and mythology?

Turtles and tortoises have historically played rich and important roles in religion and mythology in many cultures in regions where they occur as part of the native wildlife. These distinctive animals have lived on the earth for millions of years and for centuries have been seen as the companions of the gods and have been the key players in many stories of the creation of the world. Indeed, several complex myths, including the earliest noted from the ancient Chinese and the Hindus, posited that the entire universe rests on the back of a giant turtle. (What could be more important than successfully holding up the universe, and consequently the world, for millennia?) People of other cultures, including the Mongolians of East Central Asia, the Mayans of South America, and several Native North American tribes, such as the Cherokees and the Abenaki, also have myths about the world sitting on the back of a turtle or tortoise that differ only in the details. Many other Native American cultures referred to the entire North American continent as Turtle Island.

From ancient China and Japan to the Aboriginal people of Australia, and from Africa to Europe to Central and South America, myths and stories with turtles playing leading roles abound. One consistent difference in themes across cultures is that the most ancient, including the Chinese, Japanese, and Hindu, have more complex myths with turtles not only playing an active role in creation but also having roles accompanying and advising the gods and goddesses. In the Western Hemisphere, only the Mayans have elaborate ongoing roles for turtles. Most of the North American tribes have turtles' roles limited to the creation of the earth.

Stone tortoises on a cathedral in Quito, Ecuador, indicate the cultural effect that tortoises and other turtles have had on people. Photo © Judy Greene

Most of the myths and religious stories with turtles feature them in a positive, supernatural light, frequently acting as beneficent advisors to the gods. Turtles are usually viewed as having spiritual virtues and as possessing wisdom, humility, patience, prudence, constancy, and purity. Their roles are to help mankind, pursue truth, and bear the burdens of the world if not the world itself. Additional themes include contentment, old age, longevity or even immortality, reincarnated souls journeying to enlightenment, fertility, hardiness, virtue, importance, accomplishment, peace, good fortune, and success. Occasionally, however, turtles are depicted negatively as lazy, as slovenly, or as tricksters.

One of the most charming of the "world supported on the back of a turtle" myths is from the Abenaki Indians of the northeastern United States. Their version has a woman fall from her home deep into the ocean. The animals do not know what to do with her, so they call a meeting to decide and conclude that someone must swim to the depths and bring up dirt to make a home for her. Depending on the storyteller, either a toad or a muskrat finally makes it back with some soil. The turtle volunteers to have it placed on his back and eventually they bring enough to make a home for her, with the turtle as the base. The Delaware Indians also have a story in which they are all rescued by a kindly turtle. In their story, the entire tribe was saved from a great flood by piling onto the back of a huge turtle.

In some stories, the turtles are accompanied by other animals. In Japanese mythology, the God of Wisdom and Fortune is accompanied by a deer and a tortoise, and an earlier version has the tortoise accompanied by a crane, which the tortoise is sometimes believed to be married to. The

Turtles: The Animal Answer Guide

goddess identified with love, marriage, knowledge, and eloquence also is always accompanied by a tortoise.

One of the more creative of the creation stories featuring tortoises is found in the Mayan culture of Central America with a parallel version, slightly different in the details, in parts of China. In this account, the world does not sit on top of a tortoise but, rather, is inside a giant turtle with the interior surface of the turtle's carapace or shell being the sky, as if the hollow turtle were a huge version of one of the modern "snow globe" Christmas decorations that depict a winter scene. In yet another Mayan legend, the four corners of the sky, which is thought to be a huge canopy, is actually supported by an armadillo, a crab, a snail, and a turtle.

The Mayans revered turtles and tortoises more so than any other American group. Indeed the ruins of a temple that is thought to have been a shrine to all things turtle was found near Uxmal, Mexico. Many delicately carved turtles were found as decorations and were interpreted to be the theme of the temple. The Mayans were deeply religious, had a highly developed system of math, and believed the number 13 to be sacred. Their year had 13 months, and they found the 13 scutes on the carapace of a native turtle to be quite significant and meaningful, leading them to venerate turtles. The Mayans also practiced slash-and-burn agriculture. One reason they found turtles to be special was that, after burning the plot, turtles would sometimes emerge from the soil after the next heavy rain, having buried themselves to avoid the flames. This would signify the time to plant, and the Indians thought the turtles were supernatural beasts who were also messengers.

Another fascinating piece of Chinese turtle lore actually provides one explanation for how they came to have their written language of symbols. A religious ritual existed in which a priest burned the shell of a perhaps now-extinct species of turtle and then "read" and interpreted the fissures in the shell for messages. Some Chinese commentators have claimed that these fissures led to the development of written symbols and eventually the Chinese written language. Some other tidbits from turtle lore include a belief in parts of the Far East, in Japan, and maybe in parts of China, that only female turtles existed. Since there were no males in this particular myth, they were thought to have to mate with snakes to reproduce. The Chinese also thought the supernatural turtle Kwei was the longest lived of all divine beings. The Mongolians were yet another ethnic group who believed that the world existed on a tortoise's back. They placed their graves in a turtle-shaped arrangement because turtles represented immortality.

Early Buddhists believed that tortoises are souls waiting for reincarnation or enlightenment and therefore must never be killed. The Hindus, who may be the oldest society with the turtle creation myth, believed that a tortoise is the second incarnation of the God Vishna, who was transformed into a tortoise during a flood and was able to create a new world thereon. In Central Asia, one version of the origin of the world has a slightly different twist, and the creative force in this story turned the huge free-floating turtle in the giant sea onto his back and built the world on his plastron.

Are turtles depicted at all in the Christian religion?

Many people think mistakenly that turtles are mentioned in the King James version of the Old Testament in the Song of Solomon 2:12:

> For, lo! the winter is past, the rain is over and gone;
> The flowers appear on the earth;
> the time of the singing of birds is come,
> and the voice of the turtle is heard in our land.

The creature actually mentioned, while referred to as a turtle, is a turtle-dove, which is, of course, a kind of bird. It is the nature of religion and mythology that stories and parables evolve as in the example of a sweet Albanian story that weaves together Christianity and how the turtle came to have a shell. A lowly turtle wishes to comfort the Virgin Mary after the crucifixion of Christ. He is shy and covers himself with a leaf as he approaches her with great sadness. In spite of the situation, he brings a smile to her face, and from then on remains covered with a shell.

Did any early philosophers/naturalists mention turtles in their writings?

The Roman statesman-philosopher Pliny the Elder (AD 23–79), author and compiler of one of the earliest books, an encyclopedia on natural history entitled *Naturalis Historia*, wrote that turtle blood improved the eyesight and removed cataracts. He also thought it was a good treatment for venomous stings and bites from spiders and snakes and that burning tortoise flesh in one's house was a good treatment for warding off evil spirits. Aristotle (384–322 BC), the famed Greek philosopher, in his *Historia Animalia* described sea turtles rather inaccurately, attributing to them some habits that can only be those of a land tortoise. Most scholars agree that his zoological writings on topics he had studied or experienced himself were far more accurate than those that relied on information he had gathered from others.

Family: Testudinidae; Species: *Geochelone platynota;* Common Name: Burmese starred tortoise; Geographic Region: Asia

Family: Testudinidae; Species: *Geochelone platynota;* Common Name: Burmese starred tortoise; Geographic Region: Asia

Family: Testudinidae; Species: *Astrochelys radiata;* Common Name: Radiated tortoise; Geographic Region: Madagascar

Family: Testudinidae; Species: *Astrochelys radiata;* Common Name: Radiated tortoise; Geographic Region: Madagascar

Family: Emydidae; Species: *Trachemys scripta;* Common Name: Slider turtle; Geographic Region: North America

Family: Geoemydidae; Species: *Geoemyda japonica*; Common Name: Ryukyu leaf turtle; Geographic Region: Japan

Family: Geoemydidae; Species: *Geoemyda spengleri*; Common Name: Black-breasted leaf turtle; Geographic Region: Southeast Asia

Family: Emydidae; Species: *Glyptemys insculpta*; Common Name: Wood turtle; Geographic Region: North America

Family: Trionychidae; Species: *Palea steindachneri*; Common Name: Wattle-necked softshell turtle; Geographic Region: Southeast Asia

Family: Geoemydidae; Species: *Rhinoclemmys pulcherrima*; Common Name: Painted wood turtle; Geographic Region: Mexico, Central America

Family: Emydidae; Species: *Graptemys flavimaculata*; Common Name: Yellow-blotched map turtle; Geographic Region: North America

Family: Emydidae; Species: *Graptemys versa*; Common Name: Texas map turtle; Geographic Region: North America

Family: Testudinidae; Species: *Indotestudo elongata;* Common Name: Elongated tortoise; Geographic Region: Asia

Family: Geoemydidae; Species: *Leucocephalon yuwonoi;* Common Name: Sulawesi forest turtle; Geographic Region: Indonesia

Family: Kinosternidae; Species: *Sternotherus minor;* Common Name: Loggerhead musk turtle; Geographic Region: North America

Family: Chelydridae; Species: *Macrochelys temminckii;* Common Name: Alligator snapping turtle; Geographic Region: North America

Family: Emydidae; Species: *Malaclemys terrapin;* Common Name: Diamondback terrapin; Geographic Region: North America

Family: Testudinidae; Species: *Malacochersus tornieri;* Common Name: African pancake tortoise; Geographic Region: Africa

What are the roles turtles have played in children's literature?

Probably the best-known story or fable about turtles that many children learn at an early age is Aesop's fable "The Hare and the Tortoise." *Aesop's Fables* are a collection of stories, or fables, thought to have been created or compiled by a slave and storyteller in Ancient Greece named Aesop (620–560 BC). The lesson to be learned from this story is that slow, but steady, wins the race or that steady hard work will accomplish more than working in brilliant short bursts. A less-well-known Aesop's fable featured a turtle and an eagle. The turtle is on the ground complaining to his bird friends that he never gets to fly. The eagle offers to take him up and teach him. The turtle agrees and realizes too late that the eagle's plan is to drop him on the rocks and eat the resulting pieces. The lesson to us is that a person (or turtle) should consider carefully what one asks for, a lesson as important today as during Aesop's time.

Animals of all kinds are a successful and common subject in children's literature, and many popular children's books have been and are published with turtles as main characters or as supporting cast. No children's writer has been more popular nor had more success in the past few decades than Dr. Seuss, who wrote wonderful, inspiring, and educational books of nonsense rhyming verse, including *Yertle the Turtle and Other Stories*, with Yertle having the feature role in the first story.

Even Uncle Remus mentioned Brer Tarrypin in Joel Chandler Harris's book *Legends of the Old Plantation* (1881) and in the 1946 song releases of the famous movie, *Song of the South*.

Verse 4
Brother turtle and brother hare,
They ran a race and I was there.
The hare dropped in at "Barney's Place,"
And that's how the turtle won the race.

What roles do turtles play in popular culture?

Turtles periodically appear and reappear in popular culture, as anyone old enough to have experienced the Teenage Mutant Ninja Turtles (TMNT) knows. The craze began in 1984 and continued into the 1990s and beyond with movies, books, public appearances, and popular merchandise from T-shirts and Halloween costumes to coffee mugs and video games. They still have an active, almost cult, following and have a current active Web site with information on all things Ninja, including art contests, TMNT mer-

Among many cultures around the world, turtles are respected and revered, as portrayed on the country's currenc such as the sea turtle in Brazil (*left*) and on coins as with the painted terrapin (tuntung laut) in Malaysia (*right*). Phot © Margaret Wead

chandise for sale, stories, comic books, and video games to mention just a few. *Mad Magazine* parodied the fable of "The Hare and the Tortoise" in a "modern day fairy tale" in which the tortoise wins because the hare was beaten up by the Ninja turtles. That turtles have been featured in daily comic strips from Walt Kelly's turtle in *Pogo* to *B.C.* and *Over the Hedge* reveals and attests to their unending popularity with mankind. Two popular rock bands feature the name "turtle," including the 1970s band the Turtles (American) and the Mock Turtles (British), which came on the scene in the 1990s. The popular American band Red Hot Chili Peppers uses lines from Dr. Seuss's *Yertle the Turtle* as lyrics in one of their songs. The Moody Blues, the popular British band established in the 1960s and still performing into the 2000s, has a song on their album *A Question of Balance*, loosely based on "The Hare and the Tortoise."

Also, several well-known writers, including Stephen King in *It*, John Steinbeck in *The Grapes of Wrath*, Pat Conroy in *The Great Santini*, and Rita Mae Brown in one of her Sneaky Pie Brown murder mysteries, *Murder on the Prowl*, all give turtles a role. Unfortunately, the latter two make errors in their descriptions of turtles or their behavior. When Brown's cute

Turtles: The Animal Answer Guide

and courageous pet corgi, Tucker, befriends a turtle to gather clues, the beast is described as an amphibian, rather than a reptile. Conroy has the main character in his book entertain himself by running over several turtles as he drives his family from Atlanta to South Carolina during the night. Turtles in that central region of Georgia would rarely be active crossing roads on a summer night, and for someone to see several would be unheard of. However, these are minor inaccuracies that do not detract from the excellent stories.

What about turtles and math?

In mathematics and in computer science, infinite regression, more formally known as *regressus in infinitum*, is introduced by Jorge Luis Borges in his essay "Avatars of the Tortoise." Another mathematical reference to the tortoise, the cycle-finding "tortoise and the hare" algorithm, was developed in the 1960s by Robert W. Floyd.

What roles have turtles played in poetry and famous sayings?

Poets from the past and present, famous and anonymous, have written countless lines in which turtles were included, sometimes as the subject, sometimes in a subordinate role. Turtles have been featured in poems from silly to serious to sad by poets as diverse as Ogden Nash, Edward Lear, Shel Silverstein, William Shakespeare, Robert Browning, and Edna St. Vincent Millay. This is not surprising since poetry has been used as an emotional medium covering an infinite array of subjects since pen was first put to paper. Some examples of turtles in poetry are presented.

The turtle lives 'twixt plated decks
Which practically conceal its sex.
I think it clever of the turtle
In such a fix to be so fertile.

OGDEN NASH (1902–1971), "Turtle Poetry"

"'Reeling and Writhing, of course, to begin with,' the Mock Turtle replied, 'and the different branches of Arithmetic-Ambition, Distraction, Uglification, and Derision.'"

LEWIS CARROLL (Charles Lutwidge Dodgson, 1832–1898),
Alice in Wonderland, Chapter 9, "The Mock Turtle's Story"

There was an old person of Ickley,
Who could not abide to ride quickly,
He rode to Karnak on a tortoise's back,
That moony Old Person of Ickley

EDWARD LEAR (1812–1888)

Know ye the land where the cypress and myrtle
Are emblems of deeds that are done in their clime;
Where the rage of the vulture, the love of the turtle,
Now melt into sorrow, now madden to crime?
Where the virgins are soft as the roses they twine,
And all save the spirit of man, is divine?

GEORGE GORDON NOEL BYRON, LORD BYRON (1788–1824),
"The Bride of Abydos," canto i, stanza 1

I know up on the top you are seeing great sights, but down
at the bottom we, too, should have rights.

DR. SEUSS (American writer and cartoonist
best known for his collection of children's books, 1904–1991),
Yertle the Turtle and Other Stories

A plate of turtle green and glutinous

ROBERT BROWNING (1812–1889),
"The Pied Piper of Hamelin," st. iv

Have you heard of Philip Slingsby,
Slingsby of the manly chest:
How he slew the Snapping Turtle
In the regions of the west?

SIR ROBERT AYTOUN (1570–1638),
"The Fight with the Snapping Turtle"

Come from the hole where the dark days drew thee,
Wake, Methuselah! Wag thy tail!
Sniff the snare of the winds that woo thee,
Sun-kissed cabbage and sea-blown kale.
To the salted breath of the sea-bear's grot
And the low sweet laugh of the hippopot
Wake, for thy devotees can't undo thee
To see if thou really art live and hale.
Leap to life, as the leaping squirrel
Flies in fear of the squirming skink;

Gladden the heart of the keeper, Tyrrell;
Give Mr. Pocock a friendly wink!
Flap thy flippers, O thou most fleet
As once in joyance of things to eat;
Bid us note that thou still art virile
And not imbibing at Lethe's brink.
Art thou sleeping, and wilt thou waken?
Hast thou passed to the Great Beyond,
Where the Great Auk and the cavernous Kraken
Frisk and footle with all things fond;
Where the Dodo fowl and the great Dinornis
Roost with the Roc and the Aepyornis,
Where the dew drips down from the fern tree shaken
As the pismire patters through flower and frond?
Art thou sleeping, adream of orgies
In sandy coves of the Seychelles isles,
Or where in warm Galapagos gorges
The ocean echoes for miles and miles?
Of sun-warmed wastes where the wind sonorous
Roared again to thy full-mouthed chorus,
Far from bibulous Bills and Georges
That smack thee rudely with ribald smiles.
Dost thou dream how, a trifling tortoise,
The hot sun hatched thee in shifting sand,
Before the wrongs that the Roundheads brought us
Set Oliver Cromwell to rule the land?
Of an early courtship, when Pym and his earls
Were making things lively for good King Charles?
Not one left of them! Exit sortis
(Horace), but thou art still on hand.
Grant, thou monarch of eld, a token
Of blood now fired with the breath of Spring;
For the crowbar's bent and the pickaxe broken
With which we endeavored to "knock and ring."
At the warm love-thrill of the Spring's behest
That biddeth the mating bird to nest,
Wake to the word that the wind bath spoken,
Wake, old sportsman, and have thy fling!

AUTHOR UNKNOWN, "Wake, Methuselah!
(Tortoise in Midwinter at the London Zoo)"

Comment from Peter C. H. Pritchard regarding "Wake, Methuselah!":
While cleaning out my attic, I found this poem that I laboriously copied out in the library of my boarding school in Ireland about 40 years ago. It came from a bound volume of *Punch* (the British humorous weekly magazine), decades old, probably from around the turn of the century. Parts of the poem suggest some uncertainty as to whether the animal in question was an Indian Ocean or a Galápagos tortoise. Indeed, up to about the mid-nineteenth century, it should be remembered that it was common to call all giant tortoises *Testudo indica*. Certain phrases (flap thy flippers, sandy coves of the Seychelles Isles, hot sun hatched thee in shifting sands) also suggest confusion between giant tortoises and sea turtles. But no matter, the poem is magnificent. Published in *Punch* Magazine, ca. 1900, original title and author unknown.

I pulled my car aside today, to watch a trailer pass,
The neatest little trailer job, compact in line and mass,
Without an inch of wasted space within its nifty frame.
It had no car to pull it but it got there just the same.
So perfectly designed it was, to fit the driver's need,
It didn't lack a single thing except it hadn't speed.
The driver was an awful dub, he didn't seem to know
The traffic rules or when to stop or where he ought to go.
He went right through a Stop-sign on the wrong side of the road.
He didn't see the great big truck with overburdened load
Come whamming down the highway like a fearful juggernaut.
He heard the roar but not in time to keep from getting caught.
………………………………..
These dotted lines are kinder than some vivid words to show
What happened to the trailer, compact and neat . . . but slow.
Some mangled flesh, some bits of shell were wreckage to explain
Why this dusty little turtle will not cross a road again.

Don Blanding,
"Tragedy of the Road," published in
Chelonian Conservation and Biology (2000)

Anders G. J. Rhodin, editor and publisher of Chelonian Conservation and Biology, *offered the following comment regarding "Tragedy of the Road":*
For this special focus issue I had hoped to find a poem on the subject of Blanding's turtles, but searched in vain. Instead, by good fortune, I came across this light little piece authored by Don Blanding in 1946. One wonders whether this modern nature observer might not be a relative or direct descendant of William Blanding, the original collector and first

observer of Blanding's turtle back in 1838. What better way, perhaps, to honor the turtle than to present a poem by a bearer of the patronym's name. The temporal continuity from one Blanding in 1838 to another in 1946 brings a certain sense of circularity to man's observations of turtles over time. Our observations of turtles lead to an ever-increasing body of knowledge, concern, passion, and hope for the future, as those observations lead to levels of knowledge on several planes, both scientific and personal, tied together into the fabric of human chelonian experience.

He gave the solid rail a hateful kick.
From far away there came an answering tick,
And then another tick. He knew the code:
His hate had roused an engine up the road.
He wished when he had had the track alone
He had attacked it with a club or stone
And bent some rail wide open like a switch,
So as to wreck the engine in the ditch.
Too late though, now, he had himself to thank.
Its click was rising to a nearer clank.
Here it came breasting like a horse in skirts.
(He stood well back for fear of scalding squirts.)
Then for a moment all there was was size,
Confusion, and a roar that drowned the cries
He raised against the gods in the machine.
Then once again the sandbank lay serene.
The traveler's eye picked up a turtle trail,
Between the dotted feet a streak of tail,
And followed it to where he made out vague
But certain signs of buried turtle's egg;
And probing with one finger not too rough,
He found suspicious sand, and sure enough,
The pocket of a little turtle mine.
If there was one egg in it there were nine,
Torpedo-like, with shell of gritty leather,
All packed in sand to wait the trump together.
"You'd better not disturb me anymore,"
He told the distance, "I am armed for war.
The next machine that has the power to pass
Will get this plasm in its goggle glass."

                    ROBERT FROST, "The Egg and the Machine"

Copyright © 1930 by Holt, Rinehart and Winston, Inc. First published as "The Walker" in: Kreymborg, A., Mumford, L., and Rosenfeld, P. (Eds.). 1928. *The Second*

American Caravan. New York. Republished in: Frost, Robert. 1930. *Collected Poems*. New York: Henry Holt. Reprinted in: Lathem, E. C. (Ed.). 1979. *The Poetry of Robert Frost: The Collected Poems. Complete and Unabridged*. New York: Henry Holt and Co., pp. 269–270.

Let the bird of loudest lay,
On the sole Arabian tree,
Herald sad and trumpet be,
To whose sound chaste wings obey.
But thou shrieking harbinger,
Foul precurrer of the fiend,
Augur of the fever's end,
To this troop come thou not near!
From this session interdict
Every fowl of tyrant wing,
Save the eagle, feathered king:
Keep the obsequy so strict.
Let the priest in surplice white,
That defunctive music can,
Be the death-divining swan,
Lest the requiem lack his right.
And thou treble-dated crow,
That thy sable gender mak'st
With the breath thou giv'st and tak'st,
'Mongst our mourners shalt thou go.
Here the ant-hem doth commence:
Love and constancy is dead,
Phoenix and the turtle fled
In a mutual flame from hence.
So they loved as love in twain
Had the essence but in one;
Two distincts, division none;
Number there in love was slain.
Hearts remote, yet not asunder;
Distance, and no space was seen
'Twixt this turtle and his queen;
But in them it were a wonder.
So between them love did shine
That the turtle saw his right
Flaming in the phoenix' sight;
Either was the other's mine.
Property was thus appalled,

That the self was not the same;
Single nature's double name
Neither two nor one was called.
Reason, in itself confounded,
Saw division grow together,
To themselves yet either neither;
Simple were so well compounded;
That it cried, "How true a twain
Seemeth this concordant one!
Love hath reason, reason none,
If what parts can so remain."
Whereupon it made this threne
To the phoenix and the dove,
Co-supremes and stars of love,
As chorus to their tragic scene.
 Threnos
Beauty, truth, and rarity
Grace in all simplicity,
Here enclosed in cinders lie.
Death is now the phoenix' nest;
And the turtle's loyal breast
To eternity doth rest,
Leaving no posterity
'Twas not their infirmity,
It was married chastity.
Truth may seem, but cannot be;
Beauty brag, but 'tis not she:
Truth and Beauty buried be.
To this urn let those repair
That are either true or fair;
For these dead birds sigh a prayer.

WILLIAM SHAKESPEARE, "The Phoenix and the Turtle"

I
On the Coast of Coromandel
Where the early pumpkins blow,
In the middle of the woods
Lived the Yonghy-Bonghy-Bò.
Two old chairs, and half a candle,—
One old jug without a handle,—
These were all his worldly goods:
In the middle of the woods,

These were all the worldly goods,
Of the Yonghy-Bonghy-Bò,
Of the Yonghy-Bonghy-Bò.

II
Once, among the Bong-trees walking
Where the early pumpkins blow,
To a little heap of stones
Came the Yonghy-Bonghy-Bò.
There he heard a Lady talking,
To some milk-white Hens of Dorking,—
"Tis the lady Jingly Jones!
"On that little heap of stones
"Sits the Lady Jingly Jones!"
Said the Yonghy-Bonghy-Bò,
Said the Yonghy-Bonghy-Bò.

III
"Lady Jingly! Lady Jingly!
"Sitting where the pumpkins blow,
 "Will you come and be my wife?"
Said the Yonghy-Bonghy-Bò.
"I am tired of living singly,—
"On this coast so wild and shingly,—
"I'm a-weary of my life:
"If you'll come and be my wife,
 "Quite serene would be my life!"—
Said the Yonghy-Bonghy-Bò,
Said the Yonghy-Bonghy-Bò.

IV
"On this Coast of Coromandel,
"Shrimps and watercresses grow,
"Prawns are plentiful and cheap,"
Said the Yonghy-Bonghy-Bò.
"You shall have my chairs and candle,
 "And my jug without a handle!—
"Gaze upon the rolling deep
("Fish is plentiful and cheap)
"As the sea, my love is deep!"
Said the Yonghy-Bonghy-Bò,
Said the Yonghy-Bonghy-Bò.

V

Lady Jingly answered sadly,
And her tears began to flow,—
"Your proposal comes too late,
"Mr. Yonghy-Bonghy-Bò!
"I would be your wife most gladly!"
(Here she twirled her fingers madly,)
 "But in England I've a mate!
"Yes! you've asked me far too late,
"For in England I've a mate,
"Mr. Yonghy-Bonghy-Bò!
"Mr. Yonghy-Bonghy-Bò!"

VI

"Mr. Jones—(his name is Handel,—
"Handel Jones, Esquire, & Co.)
Dorking fowls delights to send,
"Mr. Yonghy-Bonghy-Bò!
"Keep, oh! keep your chairs and candle,
"And your jug without a handle,—
"I can merely be your friend!
"—Should my Jones more Dorkings send,
"I will give you three, my friend!
"Mr. Yonghy-Bonghy-Bò!
"Mr. Yonghy-Bonghy-Bò!"

VII

"Though you've such a tiny body,
"And your head so large doth grow,—
"Though your hat may blow away,
"Mr. Yonghy-Bonghy-Bò!
"Though you're such a Hoddy Doddy—
"Yet a wish that I could modi—
"fy the words I needs must say!
"Will you please to go away?
"That is all I have to say—
"Mr. Yonghy-Bonghy-Bò!
"Mr. Yonghy-Bonghy-Bò!"

VIII

Down the slippery slopes of Myrtle,
Where the early pumpkins blow,

To the calm and silent sea
Fled the Yonghy-Bonghy-Bò.
There, beyond the Bay of Gurtle,
Lay a large and lively Turtle,—
"You're the Cove," he said, "for me
"On your back beyond the sea,
"Turtle, you shall carry me!"
Said the Yonghy-Bonghy-Bò,
Said the Yonghy-Bonghy-Bò.

IX

Through the silent-roaring ocean
Did the Turtle swiftly go;
Holding fast upon his shell
Rode the Yonghy-Bonghy-Bò.
With a sad primeval motion
Towards the sunset isles of Boshen
Still the Turtle bore him well.
Holding fast upon his shell,
"Lady Jingly Jones, farewell!"
Sang the Yonghy-Bonghy-Bò,
Sang the Yonghy-Bonghy-Bò.

X

From the Coast of Coromandel,
Did that Lady never go;
On that heap of stones she mourns
For the Yonghy-Bonghy-Bò.
On that Coast of Coromandel,
In his jug without a handle
Still she weeps, and daily moans;
On that little heap of stones
To her Dorking Hens she moans,
For the Yonghy-Bonghy-Bò,
For the Yonghy-Bonghy-Bò.

EDWARD LEAR, "The Courtship of the Yonghy-Bonghy-Bò"

Cats is "dogs" and rabbits is "dogs" and so's Parrots,
but this 'ere "Tortis" is a insect, and there ain't no charge for it.

Punch, 1869, vol. 57, p. 96.

"And how many hours a day did you do lessons?" said Alice, in a hurry to change the subject. "Ten hours the first day," said the Mock Turtle: "nine the next and so on."

"What a curious plan!" exclaimed Alice.

"That's the reason they're called lessons," the Gryphon remarked: "because they lessen from day to day."

Lewis Carroll (1832–1898),
Alice's Adventures in Wonderland, chapter 9

Behold the turtle. He makes progress only when he sticks his neck out.

James Bryant Conant (1893–1978)

A turtle travels only when it sticks its neck out.   Korean Proverb

Slow and steady wins the race.

Aesop (620–560 BC), "The Hare and the Tortoise"

The two are unrelated. I'm not into turtles or space stuff.

Harry Connick, Jr., musician, singer, and actor,
explaining the title of his new album, *Star Turtle*, 1996

Parrots, tortoises and redwoods live a longer life than men do; Men a longer life than dogs do; Dogs a longer life than love does.

Edna St. Vincent Millay, poet and playwright

"Tortoises are cynics, they always expect the worst."
   "Why?"
   "I dunno, cos it often happens to them I suppose."

Terry Pratchett, author

You're slower than a herd of turtles stampeding through peanut butter.

Dilbert's "Words of Wisdom"

A well-known scientist (some say it was Bertrand Russell) once gave a public lecture on astronomy. He described how the earth orbits around the sun and how the sun, in turn, orbits around the center of a vast collection of stars called our galaxy. At the end of the lecture a little old lady at the back of the room got up and said: "What you have told us is rubbish. The world is really a flat plate supported on the back of a giant tortoise." The scientist gave a superior smile before replying, "What is the turtle standing on?" "You're very clever, young man, very clever," said the little old lady. "But it's turtles all the way down."

STEPHEN WILLIAM HAWKING (b. 1942), British theoretical physicist

All the thoughts of a turtle are turtle.

RALPH WALDO EMERSON (1803–1882),
American poet, lecturer, and essayist

It does not affect your daily life very much if your neighbor marries a box turtle. But that does not mean it is right. Now you must raise your children up in a world where that union of man and box turtle is on the same legal footing as man and wife. JOHN CORNYN, U.S. senator from Texas

Nature does not forget beauty of outline even in a mud turtle's shell.

HENRY DAVID THOREAU

Try to be like the turtle—at ease in your own shell. BILL COPELAND

Anytime you see a turtle up on top of a fence post, you know he had some help. ALEX HALEY

Looking for peace is like looking for a turtle with a mustache: You won't be able to find it. But when your heart is ready, peace will come looking for you. AJAHN CHAH (1918–1992), famous Thai Buddhist

It's OK to do cute little things like kissing a turtle, but you can't kiss another person because he's a different color? Give me a break. And you have to remember, I'm from Dallas, Texas. AARON SPELLING

Long legs mean nothing to a turtle. JAPANESE PROVERB

The tortoise does not have much to give, but it knows how to care for its child. ASHANTI PROVERB

A turtle lays a thousand eggs without anyone's knowing it; a hen lays one egg and the whole town hears. MALAY PROVERB

A piece of wood always floats near a blind tortoise just when it is in need of help. JAPANESE PROVERB

Just because he is round does not mean that the snapping turtle should be compared to the moon.   JAPANESE PROVERB

Even a tortoise can climb a fallen tree.   MALAYSIAN PROVERB

When a tortoise burns himself, he keeps his pain to himself.
                                                            CHINESE PROVERB

Turtles in Stories and Literature   127

Chapter 12

"Turtleology"

Who studies turtles?

Many kinds of people are interested in learning about turtles. Turtle biologists conduct research to determine the basic ecology, behavior, and other aspects of the biology of turtles. Conservation biologists study turtle ecology in particular, including such factors as when turtles lay eggs, how many they lay, and what the sources of mortality in a population are. This allows land and habitat managers to develop programs that are in the best conservation interests of turtles. The scientific study of turtles, especially in regard to promoting their conservation, received a considerable boost in 1992 with the founding of the Chelonian Research Foundation by Dr. Anders Rhodin and the publication of the scientific journal *Chelonian Conservation and Biology*.

Zookeepers also must learn all they can about the biology of turtles kept in captivity. What do they eat? How much sunlight do they need? How can they be induced to lay eggs? All of these questions must be answered by caretakers at zoos if they are to maintain a reproducing population of turtles. Turtle hobbyists and dealers in the pet trade also study turtles both as a matter of interest but also for purposes of knowing how best to care for the different species. Finally, some individuals simply enjoy learning more about different groups of organisms and therefore study what they can find about turtles through reading books or observing them in nature.

A first step in most ecological studies of turtles is to catch the animals. Seining is a technique commonly used by ichthyologists to catch fish but is also effective for capturing turtles in open waters. One of the best uses of seining to collect turtles has been in the tidal creeks around Kiawah Island, South Carolina, where hundreds of terrapins have been captured during long-term population studies. *Top*, Photo © Cris Hagen; *bottom*, SREL file photo

Which species are best known?

Among the species of freshwater turtles that are best known to the scientific community and the general public are the painted turtle (*Chrysemys picta*), slider turtle (*Trachemys scripta*), common box turtle (*Terrapene carolina*) and ornate box turtle (*Terrapene ornata*), snapping turtle (*Chelydra serpentina*), and some of the sea turtles. Sea turtles are popularly known and certain aspects of their natural history, such as nesting ecology, have been thoroughly studied. The giant tortoises of the Galápagos are popularly recognized but have not been as thoroughly studied as the other species mentioned.

Overall, the slider turtle is probably the most widely known species in the world because of the extensive ecological research that has been conducted on the species and its widespread popularity and distribution: in the pet trade it is known by the subspecies known as the red-eared slider (*Trachemys scripta elegans*). It is also the most commonly sold turtle species in the pet trade. The extent of the establishment of red-eared sliders as an introduced species in many countries, including England, Italy, and Japan, and even an out-of-range population in southern Florida, is profound.

The drift fence with pitfall traps is one of the most effective techniques used by turtle research biologists for gathering information on terrestrial movement patterns of turtles around wetland habitats. Among the valuable data provided are when and where turtles nest, when they move to hibernation sites, and when hatchlings leave the nest and enter the water. *Top*, SREL file photo; *bottom*, Photo © Whit Gibbons

Painted turtles are particularly well known in the northeastern and midwestern United States because they are frequently seen in and around lakes and impoundments, and the species is the most obvious turtle encountered crossing roads and basking on logs. Another turtle that is especially well-known in North America is the box turtle. Two species, the common box turtle of the eastern United States and the ornate box turtle of the Midwest and Southwest, have been encountered by almost anyone who has spent time driving in rural areas of the country within the range of one or both species. Box turtles are noted for their propensity to cross highways, and literally thousands have been picked up and taken home as pets.

On the basis of the number and diversity of scientific publications, the slider turtle and painted turtle have been the most extensively studied turtles in the world in terms of overall life history, population ecology, and

An enjoyable way to capture turtles for a research project is to catch them in the water by hand. Photo © David Scott

basic natural history. Scientific information about the slider turtle and related species was first chronicled through extensive research by Fred Cagle of Tulane University. The most extensive long-term population studies of yellow-bellied slider turtles have been conducted on the Savannah River Site in South Carolina by students, faculty, technicians, and visiting scientists.

The first field studies began in 1967, and many investigators have provided insights into the biology of the species. Included among the published works are numerous investigators, many of whom are indicated with their academic or other affiliation in the early 2000s noted in parentheses. David Clark (Office of Tropical Studies, La Selva, Costa Rica), Bob Parmenter (University of New Mexico), and Hal Avery (Drexel University) investigated selected aspects of the diet and feeding preferences of the slider turtles. David Scott (Savannah River Ecology Laboratory [SREL], University of Georgia), Tom Hinton (French Institute of Radiation and Nuclear Safety), Eric Peters (Chicago State University), Trip Lamb (East Carolina University), and John Bickham (Texas A&M University) examined various responses to exposure to ionizing radiation, including genetic effects. Joe Schubauer (U.S. Environmental Protection Agency), Jim Spotila (Drexel University), and Ray Semlitsch (University of Missouri) investigated the thermal ecology of the species in a reservoir receiving thermally elevated cooling waters from reactors. Justin Congdon (SREL, University of Georgia), Steve Morreale (Cornell University), and Vincent Burke (Johns Hopkins University Press) assessed movement patterns of the species. Nat Frazer (Utah State University) used the SREL slider turtle data to assess

Slider turtles (*Trachemys scripta*) are one of the best-known species of turtles in the world among turtle research biologists and in the pet trade. The species is commonly seen basking at the edges of wetlands throughout much of the southeastern United States. Photo © David Scott

survivorship and longevity of the species. Rebecca Yeomans (College of Coastal Georgia) studied orientation in the species. Others involved in the SREL studies on this species included Michael H. Smith (SREL, University of Georgia), Kim Scribner (Michigan State University), Kurt Buhlmann (SREL, University of Georgia), Rich Seigel (Towson University), Tracey Tuberville (SREL, University of Georgia), I. Lehr Brisbin (SREL, University of Georgia), Russ Bodie (University of Missouri), Gerald Esch (Wake Forest University), Joe Bourque (Martinez, CA), Dave Marcogliese (St. Lawrence Centre, Montreal, Quebec), Tony Tucker (Mote Marine Laboratory), Herman Eure (Wake Forest University), and Tim Goater (Malaspina University-College, Nanaimo, British Columbia).

In addition, numerous studies of the species that have been conducted by researchers in other parts of the range of the species have augmented understanding of its biology through publication of a variety of population and ecology studies in different locations within the natural range of the species. Some of the many contributions include those by John Turner, Joe Mitchell, Carl Ernst, Bill Gartska, Bob Gatten, Dan Gist, Ed Moll, Don Moll, Ed Standora, Jim Berry, Jim Christiansen, John Legler, John Tucker, Francis Rose, Kurt Buhlmann, Mike Seidel, Rich Seigel, and Steve Gotte. The biology of the species was summarized in depth in *Life History*

Before the advent of digital camera technology, turtle biologists used photocopying to keep "fingerprint" records (configuration of the seams on the shell, pattern of spotting, shell scarring or other blemishes) of individual turtles so that identification could be confirmed for those captured as many as two or three decades later. Photo © Whit Gibbons

and Ecology of the Slider Turtle (Smithsonian Institution Press), in which 41 of the world's leading turtle biologists contributed what they knew about all biological aspects of the species, including taxonomy, geographic distribution, behavior, physiology, bioenergetics, reproduction, growth, population ecology, demography, life history, and genetics.

The ecology of painted turtles is particularly well known because of extensively studied populations in Pennsylvania, Michigan, Virginia, and other regions. Owen Sexton, Henry Wilbur, Carl Ernst, and Whit Gibbons conducted independent studies in the 1950s (Sexton) and 1960s (Wilbur, Ernst, and Gibbons) that addressed growth rates, demography, movement patterns, general population ecology, and other aspects of their biology. The classic studies by Justin Congdon (SREL) and colleagues at the University of Michigan's E. S. George Reserve on aging phenomena, nesting patterns, hatchling survival, and reproductive ecology of painted turtles provided biological information unprecedented for any species of turtle. As with slider turtles, numerous regional studies have been conducted that expand the biological information base for the species. Nat Frazer used the long-term data from Sherriff's Marsh in southern Michigan to develop growth rate models, offering one of the earliest suggestions that growth in the species was being affected by global warming. Many others have made regional contributions to understanding the biology of the species as well.

In the pet trade, determining which species of turtles are best known would be based on different criteria than which ones have been most thoroughly studied or are most obvious to the general public. The popularity of species changes over time, but the list of best-known turtles for pets are several U.S. species: red-eared sliders (*Trachemys scripta*), river cooters (*Pseudemys concinna*), Florida cooters (*Pseudemys floridana*), northern red-bellied cooters (*Pseudemys rubriventris*), Florida red-bellied cooters (*Pseudemys nelsoni*), spotted turtles (*Clemmys guttata*), loggerhead musk turtles (*Sternotherus minor*), razor-backed musk turtles (*Sternotherus carinatus*), Mississippi map turtles (a subspecies of the false map turtle, *Graptemys pseudogeographica kohnii*), Ouachita map turtles (*Graptemys ouachitensis*), Florida softshells (*Apalone ferox*), common box turtles, ornate box turtles, and those from other continents: African spurred tortoises (*Centrochelys sulcata*), Horsfield's tortoise (also called Russian tortoise in the pet trade; *Agrionemys horsfieldii*), leopard tortoises (*Stigmochelys pardalis*), redfoot tortoises (*Geochelone carbonaria*), Chinese stripe-neck turtles (*Ocadia sinensis*), Southeast Asian Asian box turtles (*Cuora amboinensis*), Reeve's turtles (*Chinemys reevesei*), and painted wood turtles (*Rhinoclemmys pulcherimma manni*). Although other species are sometimes sold or traded, these have continued to be some of the most dominant species in the pet trade during the latter half of the twentieth century and the beginning of the twenty-first century.

Which species are least known?

A few species of turtles have been studied extensively in various parts of their geographic range; however, most species have been examined biologically for only certain facets of their life or in only one or two locations. Although the general geographic distribution is known for all species and basic details of nesting periods, diet, and habitat are known for most, in-depth ecological and behavioral studies have not been conducted on the majority of the world's species. Even in the United States where long-term, intensive studies have been conducted on many species, a thorough understanding of the life history and ecology of some species (e.g., striped mud turtle [*Kinosternon baurii*], Florida red-bellied cooter, razor-backed musk turtle) is still forthcoming. A few species of turtles are so rare or located in such remote locations that practically nothing is known of their general ecology, behavior, or the details of most aspects of their biology. The least-known species of all are arguably McCord's box turtle (*Cuora mccordi*) and Zhou's box turtle (*Cuora zhoui*) in which the first specimens known to science were found in markets in China, where previously no turtle biologists had awareness of their existence. The exact geographic distribution in the wild is speculated on but not precisely known for either species and the ecology of neither has been investigated.

How do scientists tell turtles apart?

The different species of turtles can be distinguished from one another by a variety of characteristics, including the shape of the shell, positions of

Although a common turtle in the clear spring runs in Florida and other streams of the southeastern United States, many aspects of the reproductive patterns and other aspects of the ecology of the loggerhead musk turtle are poorly known. Photo © Cris Hagen

the scutes on the carapace and plastron, their body size, the structure of the legs and feet, color patterns, the geographic region from which the turtle came, and sometimes by the habitat occupied by the turtle. The basic shell structure alone is often enough to distinguish between some species. For example, the green sea turtle (*Chelonia mydas*) and loggerhead sea turtle (*Caretta caretta*) along the Atlantic Coast can be distinguished from one another by whether the plates running alongside the middle of the carapace touch the first scute at the front of the turtle (loggerhead). The leatherback sea turtle (*Dermochelys coriacea*) can be distinguished from either because of the texture of the leathery shell compared with the hard shell of other sea turtles. With few exceptions, the high-domed shell characteristic of most tortoises readily distinguishes them from the more compressed and flatter shells of aquatic species. The flat leathery shells of softshell turtles separate members of that family of turtles from others that are found in the same areas.

Telling species of turtles apart in areas of high biodiversity where several species similar in appearance live in the same habitat can be more difficult. The elephantine feet of tortoises that are highly functional on land are readily distinguishable from many of the fully aquatic species, and marine turtles have flippers that distinguish them from other species. But for closely related species within the same genus or even the same family, identification can be difficult. Some of the key characteristics used to differentiate between some of the freshwater turtles are whether the head and limbs have stripes or spots, which most often are yellow, and the configuration of markings. The southern painted turtle (*Chrysemys picta dorsalis*) has a red stripe down the center of a black shell, making it readily distinguishable from any other of the southeastern aquatic species. For some species, more

Exposing children to turtles at a young age is an educational tool for developing public appreciation for turtle conservation. Photo © Mike Gibbons

subtle forms of identification are sometimes used, such as whether the upper jaw has a notch in the center of it, as with the Florida cooter (*Pseudemys floridana*), or has a sharp beak, as with the snapping turtle (*Chelydra serpentina*) and alligator snapping turtle (*Macrochelys temminckii*). Many turtle biologists must turn to field guides with keys and descriptions to distinguish between some species using a combination of characters, none of which is 100 percent flawless for identification.

One of the classic examples of the use of a combination of shell measurements was that of Trip Lamb at SREL who confirmed an earlier observation by Mike Duever that the striped mud turtle (*Kinosternon baurii*), known only from Florida and southern Georgia, actually occurred on the Savannah River Site in South Carolina. The presence of the striped mud turtle was not known formerly because individuals in South Carolina do not normally have the three yellow stripes on the carapace that are typical of specimens from Florida, although they do have stripes on the head. In a follow-up study, Trip Lamb and Jeffrey Lovich actually demonstrated that the striped mud turtle was much more widespread, ranging as far as Virginia in the Atlantic Coastal Plain.

Another approach used by scientists not only for identifying turtles but for determining the phylogenetic relationships among species is the use of molecular genetics. Molecular techniques such as DNA analyses are often used to distinguish turtle species from one another that are very similar in physical appearance. Despite the revelation in recent decades of genetic and phylogenetic relationships among some animals, including turtles that have been unsuspected, DNA studies have not altered many of the classical relationships among modern turtles that were established by early systematists based on shell and bone structures.

Scientific and Common Names of Living Turtles

Not all turtle authorities agree on the scientific or common names that should be applied to particular species. The following taxonomic classification of families, genera, and species is structured after the IUCN "Turtles of the World" checklist (Rhodin, Van Dijk, and Parham 2008), with modifications based on Ernst and Barbour (1989); Ernst, Lovich, and Barbour (1994); Bonin, Devaux, and Dupré (2006); and Buhlmann, Tuberville, and Gibbons (2008); and Ernst and Lovich (2009).

Number of families: 14
Number of genera: 104
Number of species: 318

Alphabetical Listing of Turtle Genera within Families

SUBCLASS PLEURODIRA

Family	Genus
Chelidae	*Acanthochelys*
	Chelodina
	Chelus
	Elseya
	Elusor
	Emydura
	Hydromedusa
	Phrynops
	Platemys
	Pseudemydura
	Rheodytes
Pelomedusidae	*Pelomedusa*
	Pelusios
Podocnemidae	*Erymnochelys*
	Peltocephalus
	Podocnemis

SUBCLASS CRYPTODIRA

Family	Genus
Chelydridae	*Chelydra*
	Macrochelys
Platysternidae	*Platysternon*
Cheloniidae	*Caretta*
	Chelonia
	Eretmochelys
	Lepidochelys
	Natator
Dermochelyidae	*Dermochelys*
Carettochelyidae	*Carettochelys*
Trionychidae	*Amyda*
	Apalone
	Aspideretes
	Chitra
	Cyclanorbis
	Cycloderma
	Dogania
	Lissemys

Family	Genus	Family	Genus
	Nilssonia		*Geoemyda*
	Palea		*Hardella*
	Pelochelys		*Heosemys*
	Pelodiscus		*Hieremys*
	Rafetus		*Kachuga*
	Trionyx		*Malayemys*
Dermatemydidae	*Dermatemys*		*Mauremys*
Kinosternidae	*Claudius*		*Melanochelys*
	Kinosternon		*Morenia*
	Staurotypus		*Notochelys*
	Sternotherus		*Ocadia*
Testudinidae	*Chersina*		*Pyxidea*
	Geochelone		*Rhinoclemmys*
	Gopherus		*Sacalia*
	Homopus		*Siebenrockiella*
	Indotestudo	Geoemydidae	*Agrionemys*
	Kinixys		*Astrochelys*
	Malacochersus		*Batrachemys*
	Manouria		*Bufocephala*
	Psammobates		*Centrochelys*
	Pyxis		*Chelonoidis*
	Testudo		*Cistoclemmys*
Emydidae	*Actinemys*		*Dipsochelys*
	Chrysemys		*Leucocephalon*
	Clemmys		*Mesoclemmys*
	Deirochelys		*Oscaria*
	Emydoidea		*Pangshura*
	Emys		*Ranacephala*
	Glyptemys		*Rhinemys*
	Graptemys		*Stigmochelys*
	Malaclemys		
	Pseudemys		
	Terrapene		
	Trachemys		
Bataguridae	*Batagur*		
	Callagur		
	Chinemys		
	Cuora		
	Cyclemys		
	Geoclemys		

Alphabetical Listing of All Living Turtles by Genus and Species

Family	Species	Common Name	General Location
Chelidae	*Acanthochelys macrocephala*	Pantanal swamp turtle	South America
Chelidae	*Acanthochelys pallidipectoris*	Chaco swamp turtle	South America
Chelidae	*Acanthochelys radiolata*	Coastal swamp turtle	South America
Chelidae	*Acanthochelys spixii*	Spix's swamp turtle	South America
Emydidae	*Actinemys marmorata*	Pacific pond turtle	North America
Testudinidae	*Agrionemys horsfieldii*	Horsfield's tortoise	Asia
Testudinidae	*Aldabrachelys arnoldi*	Arnold's tortoise	Seychelles Islands
Testudinidae	*Aldabrachelys dussumieri*	Giant Aldabra tortoise	Seychelles Islands
Testudinidae	*Aldabrachelys hololissa*	Hololissa tortoise	Seychelles Islands
Trionychidae	*Amyda cartilaginea*	Asiatic softshell turtle	Southeast Asia, India
Trionychidae	*Apalone ferox*	Florida softshell turtle	North America
Trionychidae	*Apalone mutica*	Smooth softshell turtle	North America
Trionychidae	*Apalone spinifera*	Spiny softshell turtle	North America
Trionychidae	*Aspideretes gangeticus*	Indian softshell turtle	India
Trionychidae	*Aspideretes hurum*	Indian peacock softshell turtle	India, Nepal
Trionychidae	*Aspideretes leithii*	Leith's softshell turtle	India
Trionychidae	*Aspideretes nigricans*	Black softshell turtle	India, Bangladesh
Testudinidae	*Astrochelys radiata*	Radiated tortoise	Madagascar
Testudinidae	*Astrochelys yniphora*	Ploughshare tortoise	Madagascar
Geoemydidae	*Batagur affinis*	Southern mangrove terrapin	Southeast Asia
Geoemydidae	*Batagur baska*	Mangrove terrapin	India
Chelidae	*Batrachemys dahli*	Dahl's toad-headed turtle	South America
Chelidae	*Batrachemys heliostemma*	Yellow toad-headed turtle	South America
Chelidae	*Batrachemys nasutus*	Guianan toad-headed turtle	South America
Chelidae	*Batrachemys raniceps*	Amazon toad-headed turtle	South America
Chelidae	*Batrachemys tuberculatus*	Tuberculate toad-headed turtle	South America
Chelidae	*Batrachemys zuliae*	Zulia toad-headed turtle	South America
Chelidae	*Bufocephala vanderhaegei*	Vanderhage's toad-headed turtle	South America
Geoemydidae	*Callagur borneoensis*	Painted terrapin	Southeast Asia
Cheloniidae	*Caretta caretta*	Loggerhead sea turtle	Pantropical and temperate oceans
Carettochelyidae	*Carettochelys insculpta*	Pig-nosed turtle	Australia, New Guinea
Testudinidae	*Centrochelys sulcata*	African spurred tortoise	Africa
Chelidae	*Chelodina longicollis*	Eastern long-necked turtle	Australia
Chelidae	*Chelodina canni*	Cann's long-necked turtle	Australia
Chelidae	*Chelodina mccordi*	Rote Island long-necked turtle	Indonesia
Chelidae	*Chelodina novaeguineae*	New Guinea snake-necked turtle	New Guinea
Chelidae	*Chelodina pritchardi*	Pritchard's long-necked turtle	New Guinea
Chelidae	*Chelodina reimanni*	Reimann's snake-necked turtle	New Guinea
Chelidae	*Chelodina steindachneri*	Steindachner's snake-necked turtle	Australia
Cheloniidae	*Chelonia mydas*	Green sea turtle	Tropical to temperate seas
Testudinidae	*Chelonoidis carbonaria*	Red-footed tortoise	South America
Testudinidae	*Chelonoidis chilensis*	Patagonian tortoise	South America
Testudinidae	*Chelonoidis denticulata*	Yellowfoot tortoise	South America
Testudinidae	*Chelonoidis nigra*	Galápagos giant tortoise	South America

Family	Species	Common Name	General Location
Testudinidae	*Chelonoidis petersi*	Chaco tortoise	South America
Chelidae	*Chelus fimbriata*	Matamata	South America
Chelydridae	*Chelydra acutirostris*	South American snapping turtle	Central and South America
Chelydridae	*Chelydra rossignonii*	Central American snapping turtle	Central America
Chelydridae	*Chelydra serpentina*	Common snapping turtle	North America
Testudinidae	*Chersina angulata*	Angulated tortoise	Africa
Geoemydidae	*Chinemys nigricans*	Red-necked pond turtle	Asia
Geoemydidae	*Chinemys reevesii*	Reeve's turtle	Asia
Trionychidae	*Chitra chitra*	SE Asian narrow-headed softshell turtle	Southeast Asia
Trionychidae	*Chitra indica*	Indian narrow-headed softshell turtle	Pakistan, India, Bangladesh
Trionychidae	*Chitra vandijki*	Burmese narrow-headed softshell turtle	Myanmar
Emydidae	*Chrysemys picta*	Painted turtle	North America
Kinosternidae	*Claudius angustatus*	Narrow-bridged turtle	Mexico, Central America
Emydidae	*Clemmys guttata*	Spotted turtle	North America
Geoemydidae	*Cuora amboinensis*	SE Asian box turtle	Southeast Asia
Geoemydidae	*Cuora aurocapitata*	Yellow-headed box turtle	China
Geoemydidae	*Cuora favomarginata*	Common yellow-margined box turtle	China, Japan
Geoemydidae	*Cuora galbinifrons*	Indochinese box turtle	Southeast Asia
Geoemydidae	*Cuora mccordi*	McCord's box turtle	China
Geoemydidae	*Cuora pani*	Pan's box turtle	China
Geoemydidae	*Cuora trifasciata*	Chinese three-striped box turtle	Southeast Asia
Geoemydidae	*Cuora yunnanensis*	Yunnan box turtle	China
Geoemydidae	*Cuora zhoui*	Zhou's box turtle	China
Trionychidae	*Cyclanorbis elegans*	Nubian flap-shelled turtle	Africa
Trionychidae	*Cyclanorbis senegalensis*	Senegal flap-shelled turtle	Africa
Geoemydidae	*Cyclemys atripons*	Black-bridged leaf turtle	Southeast Asia
Geoemydidae	*Cyclemys dentata*	Asian leaf turtle	Southeast Asia
Geoemydidae	*Cyclemys enigmatica*	Enigmatic leaf turtle	Southeast Asia
Geoemydidae	*Cyclemys fusca*	Dusky leaf turtle	Southeast Asia
Geoemydidae	*Cyclemys gemeli*	Gemel leaf turtle	Southeast Asia
Geoemydidae	*Cyclemys ovata*	Borneo leaf turtle	Southeast Asia
Geoemydidae	*Cyclemys oldhami*	Oldham's leaf turtle	Asia
Trionychidae	*Cycloderma aubryi*	Aubry's flap-shelled turtle	Africa
Trionychidae	*Cycloderma frenatum*	Zambezi flap-shelled turtle	Africa
Emydidae	*Deirochelys reticularia*	Chicken turtle	North America
Dermatemydidae	*Dermatemys mawii*	Central American river turtle	Mexico, Central America
Dermochelyidae	*Dermochelys coriacea*	Leatherback sea turtle	World's oceans
Trionychidae	*Dogania subplana*	Malayan softshell turtle	Southeast Asia
Chelidae	*Elseya albagula*	Southern snapping turtle	Australia
Chelidae	*Elseya bellii*	Bell's saw-shelled turtle	Australia
Chelidae	*Elseya branderhorsti*	Southern New Guinea snapping turtle	New Guinea
Chelidae	*Elseya dentata*	Northern snapping turtle	Australia

Family	Species	Common Name	General Location
Chelidae	*Elseya georgesi*	Bellinger river turtle	Australia
Chelidae	*Elseya irwini*	Irwin's turtle	Australia
Chelidae	*Elseya jukesi*	Jukes' snapping turtle	Australia
Chelidae	*Elseya latisternum*	Saw-shelled turtle	Australia
Chelidae	*Elseya lavarackorum*	Lavarackor's snapping turtle	Australia
Chelidae	*Elseya novaeguineae*	New Guinea snapping turtle	New Guinea
Chelidae	*Elseya purvisi*	Manning River turtle	Australia
Chelidae	*Elseya schultzei*	Schultz's snapping turtle	New Guinea
Chelidae	*Elseya stirlingi*	Stirling's snapping turtle	Australia
Chelidae	*Elusor macrurus*	Mary River turtle	Australia
Emydidae	*Emydoidea blandingii*	Blanding's turtle	North America
Chelidae	*Emydura australis*	Australian big-headed turtle	Australia
Chelidae	*Emydura macquarii*	Murray River turtle	Australia
Chelidae	*Emydura subglobosa*	Red-bellied short-necked turtle	New Guinea
Chelidae	*Emydura tanybaraga*	Yellow-faced turtle	Australia
Chelidae	*Emydura victoriae*	Northern red-faced turtle	Australia
Emydidae	*Emys orbicularis*	Common European pond turtle	Europe, Africa, Middle East
Emydidae	*Emys trinacris*	Sicilian pond turtle	Europe
Cheloniidae	*Eretmochelys imbricata*	Hawksbill sea turtle	Tropical to temperate seas
Podocnemidae	*Erymnochelys madagascariensis*	Madagascar big-headed turtle	Madagascar
Testudinidae	*Geochelone elegans*	Indian star tortoise	Asia
Testudinidae	*Geochelone platynota*	Burmese starred tortoise	Asia
Geoemydidae	*Geoclemys hamiltonii*	Spotted pond turtle	Asia
Geoemydidae	*Geoemyda japonica*	Ryukyu leaf turtle	Japan
Geoemydidae	*Geoemyda spengleri*	Black-breasted leaf turtle	Southeast Asia
Emydidae	*Glyptemys insculpta*	Wood turtle	North America
Emydidae	*Glyptemys muhlenbergii*	Bog turtle	North America
Testudinidae	*Gopherus agassizii*	Desert tortoise	North America
Testudinidae	*Gopherus berlandieri*	Texas tortoise	North America
Testudinidae	*Gopherus flavomarginatus*	Mexican giant gopher tortoise	Mexico
Testudinidae	*Gopherus polyphemus*	Gopher tortoise	North America
Emydidae	*Graptemys barbouri*	Barbour's map turtle	North America
Emydidae	*Graptemys caglei*	Cagle's map turtle	North America
Emydidae	*Graptemys ernsti*	Ernst's map turtle	North America
Emydidae	*Graptemys flavimaculata*	Yellow-blotched map turtle	North America
Emydidae	*Graptemys geographica*	Common map turtle	North America, Canada
Emydidae	*Graptemys gibbonsi*	Gibbons' map turtle	North America
Emydidae	*Graptemys nigrinoda*	Black-nobbed map turtle	North America
Emydidae	*Graptemys oculifera*	Ringed map turtle	North America
Emydidae	*Graptemys ouachitensis*	Ouachita map turtle	North America
Emydidae	*Graptemys pseudogeographica*	False map turtle	North America
Emydidae	*Graptemys pulchra*	Alabama map turtle	North America
Emydidae	*Graptemys versa*	Texas map turtle	North America
Geoemydidae	*Hardella thurjii*	Crowned river turtle	Asia
Geoemydidae	*Heosemys depressa*	Arakan forest turtle	Asia

Family	Species	Common Name	General Location
Geoemydidae	*Heosemys grandis*	Giant Asian pond turtle	Southeast Asia
Geoemydidae	*Heosemys spinosa*	Spiny turtle	Southeast Asia
Geoemydidae	*Hieremys annandalii*	Yellow-headed temple turtle	Southeast Asia
Testudinidae	*Homopus areolatus*	Beaked Cape tortoise	Africa
Testudinidae	*Homopus boulengeri*	Boulenger's padloper	Africa
Testudinidae	*Homopus femoralis*	Karoo Cape tortoise	Africa
Testudinidae	*Homopus signatus*	Speckled Cape tortoise	Africa
Testudinidae	*Homopus solus*	Nama Cape tortoise	Africa
Chelidae	*Hydromedusa maximiliani*	Brazilian snake-necked turtle	South America
Chelidae	*Hydromedusa tectifera*	South American snake-necked turtle	South America
Testudinidae	*Indotestudo elongata*	Elongated tortoise	Asia
Testudinidae	*Indotestudo forstenii*	Forsten's tortoise	Indonesia
Testudinidae	*Indotestudo travancorica*	Travancore tortoise	India
Geoemydidae	*Kachuga dhongoka*	Three-striped roof turtle	Asia
Geoemydidae	*Kachuga kachuga*	Red-crowned roof turtle	Asia
Geoemydidae	*Kachuga trivittata*	Burmese roof turtle	Asia
Testudinidae	*Kinixys belliana*	Bell's eastern hinge-back tortoise	Africa
Testudinidae	*Kinixys erosa*	Serrated hinge-back tortoise	Africa
Testudinidae	*Kinixys homeana*	Home's hinge-back tortoise	Africa
Testudinidae	*Kinixys lobatsiana*	Lobatse hinge-back tortoise	Africa
Testudinidae	*Kinixys natalensis*	Natal hinge-back tortoise	Africa
Testudinidae	*Kinixys spekii*	Speke's hinge-back tortoise	Africa
Kinosternidae	*Kinosternon acutum*	Tabasco mud turtle	Mexico, Central America
Kinosternidae	*Kinosternon alamosae*	Alamos mud turtle	Mexico
Kinosternidae	*Kinosternon angustipons*	Narrow-bridged mud turtle	Central America
Kinosternidae	*Kinosternon arizonense*	Arizona mud turtle	North America
Kinosternidae	*Kinosternon baurii*	Striped mud turtle	North America
Kinosternidae	*Kinosternon chimalhuaca*	Jalisco mud turtle	Mexico
Kinosternidae	*Kinosternon creaseri*	Creaser's mud turtle	Mexico
Kinosternidae	*Kinosternon dunni*	Dunn's mud turtle	South America
Kinosternidae	*Kinosternon durangoense*	Durango mud turtle	Mexico
Kinosternidae	*Kinosternon flavescens*	Yellow mud turtle	North America, Mexico
Kinosternidae	*Kinosternon herrerai*	Herrera's mud turtle	Mexico
Kinosternidae	*Kinosternon hirtipes*	Mexican rough-footed mud turtle	North America, Mexico
Kinosternidae	*Kinosternon integrum*	Mexican mud turtle	Mexico
Kinosternidae	*Kinosternon leucostomum*	White-lipped mud turtle	Mexico, Central and South America
Kinosternidae	*Kinosternon oaxacae*	Oaxaca mud turtle	Mexico
Kinosternidae	*Kinosternon scorpioides*	Scorpion mud turtle	Mexico, Central and South America
Kinosternidae	*Kinosternon sonoriense*	Sonoran mud turtle	North America, Mexico
Kinosternidae	*Kinosternon subrubrum*	Eastern mud turtle	North America
Cheloniidae	*Lepidochelys kempii*	Kemp's ridley sea turtle	North American Atlantic Ocean and Gulf
Cheloniidae	*Lepidochelys olivacea*	Olive ridley sea turtle	South Atlantic, Pacific, Indian oceans
Geoemydidae	*Leucocephalon yuwonoi*	Sulawesi forest turtle	Indonesia

Family	Species	Common Name	General Location
Trionychidae	*Lissemys punctata*	Indian flapshell turtle	India
Trionychidae	*Lissemys scutata*	Burmese flap-shelled turtle	Asia
Chelidae	*Macrochelodina burrungandjii*	Arnhem Land snake-necked turtle	Australia
Chelidae	*Macrochelodina expansa*	Broad-shelled snake-necked turtle	Australia
Chelidae	*Macrochelodina kuchlingi*	Kuchling's long-necked turtle	Australia
Chelidae	*Macrochelodina parkeri*	Parker's long-necked turtle	New Guinea
Chelidae	*Macrochelodina rugosa*	Northern snake-necked turtle	Australia, New Guinea
Chelidae	*Macrochelodina walloyarrina*	Kimberley long-necked turtle	Australia
Chelydridae	*Macrochelys temminckii*	Alligator snapping turtle	North America
Chelidae	*Macrodiremys oblonga*	Narrow-breasted snake-necked turtle	Australia
Emydidae	*Malaclemys terrapin*	Diamondback terrapin	North America
Testudinidae	*Malacochersus tornieri*	African pancake tortoise	Africa
Geoemydidae	*Malayemys macrocephala*	Malayan snail-eating turtle	Southeast Asia
Geoemydidae	*Malayemys subtrijuga*	Mekong snail-eating turtle	Southeast Asia
Testudinidae	*Manouria emys*	Asian brown tortoise	Southeast Asia
Testudinidae	*Manouria impressa*	Impressed tortoise	Southeast Asia
Geoemydidae	*Mauremys annamensis*	Annam pond turtle	Southeast Asia
Geoemydidae	*Mauremys caspica*	Caspian terrapin	Africa, Asia
Geoemydidae	*Mauremys japonica*	Japanese pond turtle	Japan
Geoemydidae	*Mauremys leprosa*	Spanish terrapin	Europe, Africa
Geoemydidae	*Mauremys mutica*	Yellow pond turtle	Southeast Asia
Geoemydidae	*Mauremys rivulata*	Eastern Mediterranean pond turtle	Europe, Middle East
Geoemydidae	*Melanochelys tricarinata*	Tricarinate hill turtle	Asia
Geoemydidae	*Melanochelys trijuga*	Indian black turtle	Asia
Chelidae	*Mesoclemmys gibba*	Common toad-headed turtle	South America
Chelidae	*Mesoclemmys perplexa*	Brazilian toad-headed turtle	South America
Geoemydidae	*Morenia ocellata*	Burmese eyed turtle	Asia
Geoemydidae	*Morenia petersi*	Indian eyed turtle	Asia
Cheloniidae	*Natator depressus*	Flatback sea turtle	Australia
Trionychidae	*Nilssonia formosa*	Burmese peacock softshell turtle	Southeast Asia
Geoemydidae	*Notochelys platynota*	Malaysian flat-shelled turtle	Southeast Asia
Geoemydidae	*Ocadia sinensis*	Chinese stripe-necked turtle	Southeast Asia
Geoemydidae	*Orlitia borneensis*	Malayan giant river turtle	Southeast Asia
Trionychidae	*Palea steindachneri*	Wattle-necked softshell turtle	Southeast Asia
Geoemydidae	*Pangshura lentoria*	Indian tent turtle	Asia
Geoemydidae	*Pangshura smithii*	Brown roofed turtle	India
Geoemydidae	*Pangshura sylhetensis*	Assam roofed turtle	Asia
Geoemydidae	*Pangshura tecta*	Indian roofed turtle	Asia
Trionychidae	*Pelochelys bibroni*	New Guinea giant softshell turtle	New Guinea
Trionychidae	*Pelochelys cantorii*	Asian giant softshell turtle	India, Southeast Asia
Trionychidae	*Pelochelys signifera*	Webb's giant softshell turtle	New Guinea
Trionychidae	*Pelodiscus axenaria*	Hunan softshell turtle	China
Trionychidae	*Pelodiscus maackii*	Amur softshell turtle	China
Trionychidae	*Pelodiscus parviformis*	Southern Chinese softshell turtle	China
Trionychidae	*Pelodiscus sinensis*	Chinese softshell turtle	China
Pelomedusidae	*Pelomedusa subrufa*	Common African helmeted turtle	Africa
Podocnemidae	*Peltocephalus dumerilianus*	Dumeril's big-headed river turtle	South America

Family	Species	Common Name	General Location
Pelomedusidae	*Pelusios adansonii*	Adanson's mud turtle	Africa
Pelomedusidae	*Pelusios bechuanicus*	Okavango mud turtle	Africa
Pelomedusidae	*Pelusios broadleyi*	Lake Turkana mud turtle	Africa
Pelomedusidae	*Pelusios carinatus*	African keeled mud turtle	Africa
Pelomedusidae	*Pelusios castaneus*	West African mud turtle	Africa
Pelomedusidae	*Pelusios castanoides*	East African yellow-bellied mud turtle	Africa, Seychelles, Madagascar
Pelomedusidae	*Pelusios chapini*	Chapin's mud turtle	Africa
Pelomedusidae	*Pelusios cupulatta*	Cupulatta's mud turtle	Africa
Pelomedusidae	*Pelusios gabonensis*	African forest turtle	Africa
Pelomedusidae	*Pelusios marani*	Maran's sideneck turtle	Africa
Pelomedusidae	*Pelusios nanus*	African dwarf mud turtle	Africa
Pelomedusidae	*Pelusios niger*	West African black turtle	Africa
Pelomedusidae	*Pelusios rhodesianus*	Variable mud turtle	Africa
Pelomedusidae	*Pelusios sinuatus*	East African serrated mud turtle	Africa
Pelomedusidae	*Pelusios subniger*	East African black mud turtle	Africa
Pelomedusidae	*Pelusios upembae*	Upemba mud turtle	Africa, Seychelles, Madagascar
Pelomedusidae	*Pelusios williamsi*	William's mud turtle	Africa
Chelidae	*Phrynops geoffroanus*	Geoffroy's side-necked turtle	South America
Chelidae	*Phrynops hilarii*	Spot-bellied toad-headed turtle	South America
Chelidae	*Phrynops tuberosus*	Northern toad-headed turtle	South America
Chelidae	*Phrynops williamisi*	William's toad-headed turtle	South America
Chelidae	*Platemys platycephala*	Twist-necked turtle	South America
Platysternidae	*Platysternon megacephalum*	Big-headed turtle	Southeast Asia
Podocnemidae	*Podocnemis erythrocephala*	Red-headed Amazon River turtle	South America
Podocnemidae	*Podocnemis expansa*	Arrau river turtle	South America
Podocnemidae	*Podocnemis lewyana*	Magdalena River turtle	South America
Podocnemidae	*Podocnemis sextuberculata*	Six-tubercled river turtle	South America
Podocnemidae	*Podocnemis unifilis*	Yellow-spotted river turtle	South America
Podocnemidae	*Podocnemis vogli*	Savanna side-necked turtle	South America
Testudinidae	*Psammobates geometricus*	Geometric tortoise	Africa
Testudinidae	*Psammobates oculifer*	Serrated tortoise	Africa
Testudinidae	*Psammobates tentorius*	South African tent tortoise	Africa
Chelidae	*Pseudemydura umbrina*	Western swamp turtle	Australia
Emydidae	*Pseudemys alabamensis*	Alabama red-bellied cooter	North America
Emydidae	*Pseudemys concinna*	River cooter	North America
Emydidae	*Pseudemys floridana*	Pond cooter	North America
Emydidae	*Pseudemys gorzugi*	Rio Grande cooter	North America, Mexico
Emydidae	*Pseudemys nelsoni*	Florida red-bellied cooter	North America
Emydidae	*Pseudemys rubriventris*	Northern red-bellied cooter	North America
Emydidae	*Pseudemys texana*	Texas cooter	North America
Geoemydidae	*Pyxidea mouhotii*	Keeled box turtle	Southeast Asia
Testudinidae	*Pyxis arachnoides*	Common spider tortoise	Madagascar
Testudinidae	*Pyxis planicauda*	Flat-shelled spider tortoise	Madagascar
Trionychidae	*Rafetus euphraticus*	Euphrates softshell turtle	Middle East
Trionychidae	*Rafetus swinhoei*	Yangtze softshell turtle	China

Family	Species	Common Name	General Location
Chelidae	*Ranacephala hogei*	Hoge's toad-headed turtle	South America
Chelidae	*Rheodytes leukops*	Fitzroy River turtle	Australia
Chelidae	*Rhinemys rufipes*	Red toad-headed turtle	South America
Geoemydidae	*Rhinoclemmys annulata*	Brown wood turtle	Central and South America
Geoemydidae	*Rhinoclemmys areolata*	Furrowed wood turtle	Mexico, Central America
Geoemydidae	*Rhinoclemmys diademata*	Maracaibo wood turtle	South America
Geoemydidae	*Rhinoclemmys funerea*	Black wood turtle	Central America
Geoemydidae	*Rhinoclemmys melanosterna*	Colombian wood turtle	Central America, South America
Geoemydidae	*Rhinoclemmys nasuta*	Large-nose wood turtle	South America
Geoemydidae	*Rhinoclemmys pulcherrima*	Painted wood turtle	Mexico, Central America
Geoemydidae	*Rhinoclemmys punctularia*	Spot-legged wood turtle	South America
Geoemydidae	*Rhinoclemmys rubida*	Mexican spotted wood turtle	Mexico
Geoemydidae	*Sacalia bealei*	Beale's eyed-turtle	Southeast Asia
Geoemydidae	*Sacalia quadriocellata*	Four-eyed turtle	Southeast Asia
Geoemydidae	*Siebenrockiella crassicollis*	Black marsh turtle	Southeast Asia
Geoemydidae	*Siebenrockiella leytensis*	Philippine pond turtle	Philippines
Kinosternidae	*Staurotypus salvinii*	Pacific Coast giant musk turtle	Mexico, Central America
Kinosternidae	*Staurotypus triporcatus*	Mexican giant musk turtle	Mexico, Central America
Kinosternidae	*Sternotherus carinatus*	Razor-back musk turtle	North America
Kinosternidae	*Sternotherus depressus*	Flattened musk turtle	North America
Kinosternidae	*Sternotherus minor*	Loggerhead musk turtle	North America
Kinosternidae	*Sternotherus odoratus*	Common musk turtle	North America
Testudinidae	*Stigmochelys pardalis*	Leopard tortoise	Africa
Emydidae	*Terrapene carolina*	Common box turtle	North America
Emydidae	*Terrapene coahuila*	Coahuila box turtle	Mexico
Emydidae	*Terrapene nelsoni*	Southern spotted box turtle	Mexico
Emydidae	*Terrapene ornata*	Ornate box turtle	North America
Testudinidae	*Testudo graeca*	Mediterranean spur-thighed tortoise	Africa, Europe
Testudinidae	*Testudo hermanni*	Hermann's tortoise	Europe
Testudinidae	*Testudo kazachstanica*	Kazachstan steppe tortoise	Asia
Testudinidae	*Testudo kleinmanni*	Kleinmann's tortoise	Africa
Testudinidae	*Testudo marginata*	Marginated tortoise	Europe
Testudinidae	*Testudo rustamovi*	Turkmenian steppe tortoise	Asia
Emydidae	*Trachemys adiutrix*	Brazilian slider turtle	South America
Emydidae	*Trachemys callirostris*	Colombian slider turtle	South America
Emydidae	*Trachemys decorata*	Hispaniolan slider turtle	Caribbean
Emydidae	*Trachemys decussata*	Cuban slider turtle	Caribbean
Emydidae	*Trachemys dorbigni*	Orbigny's slider turtle	South America
Emydidae	*Trachemys emolli*	Nicaraguan slider turtle	Central America
Emydidae	*Trachemys gaigeae*	Plateau slider turtle	North America, Mexico
Emydidae	*Trachemys nebulosa*	Northwest Mexican slider turtle	North America, Mexico
Emydidae	*Trachemys ornata*	Ornate slider turtle	Mexico, Central and South America
Emydidae	*Trachemys scripta*	Slider turtle	North America
Emydidae	*Trachemys stejnegeri*	Central Antillean slider turtle	Caribbean

Family	Species	Common Name	General Location
Emydidae	*Trachemys taylori*	Cuatro Cienegas slider turtle	Mexico
Emydidae	*Trachemys terrapen*	Jamaican slider turtle	Caribbean
Emydidae	*Trachemys venusta*	Meso-american slider turtle	Mexico, Central America
Emydidae	*Trachemys yaquia*	Yaqui slider turtle	Mexico
Trionychidae	*Trionyx triunguis*	Nile softshell turtle	Africa, Middle East
Geoemydidae	*Vijayachelys silvatica*	Cochin cane forest turtle	India

Appendix B

Organizations and Societies
for Turtle Conservation

A diversity of regional, national, and international organizations and societies focuses on turtle conservation directly or contribute indirectly by publishing or supporting research that provides scientific data needed to make
data-based decisions. Some organizations assist turtle conservation efforts
by reporting information on conservation issues through electronic mail-
ing lists, announcements, brochures, newsletters, and the popular press.
The following list is not all inclusive but provides many of the groups that
make significant contributions to turtle conservation.

Reptile Expos, Zoos, and Public Aquariums with Turtles and Tortoises in Exhibits

Alligator Adventure, North Myrtle Beach, SC
Audubon Aquarium of the Americas, New Orleans, LA
Audubon Nature Institute (Audubon Zoo), New Orleans, LA
Bronx Zoo, Bronx, NY
Clyde Peeling's Reptiland, Allenwood, PA
Columbus Zoo and Aquarium, Columbus, OH
Detroit Zoo, Detroit, MI
Disney's Animal Kingdom, Orlando, FL
Edisto Island Serpentarium, Edisto Island, SC
Fort Worth Zoo, Fort Worth, TX
Gladys Porter Zoo, Brownsville, TX
Jacksonville Zoo, Jacksonville, FL
Kentucky Reptile Zoo, Slade, KY
Madras Crocodile Bank, Chennai, India
Miami Metrozoo, Miami, FL
Munster Zoo, Münster, Germany
National Aquarium in Baltimore, Baltimore, MD
New England Aquarium, Boston, MA
Quito Zoo, Guayllabamba, Ecuador
Reptile World Serpentarium, St. Cloud, FL
Riverbanks Zoo and Garden, Columbia, SC
San Antonio Zoo, San Antonio, TX

San Diego Zoo, San Diego, CA
Shedd Aquarium, Chicago, IL
South Carolina Aquarium, Charleston, SC
Southwick's Zoo, Mendon, MA
St. Augustine Alligator Farm, St Augustine, FL
Steinhart Aquarium, San Francisco, CA
St. Louis Zoo, St. Louis, MO
Tennessee Aquarium, Chattanooga, TN
Toronto Zoo, Toronto, Ontario, Canada
Turtleback Zoo, West Orange, NJ
Zoo Atlanta, Atlanta, GA
Zoological Society of London (ZSL) London Zoo, Regents Park, London
Zoological Society of London (ZSL) Whipsnade Zoo, Whipsnade,
    Bedfordshire, England

Scientific and Conservation Societies That Identify Turtles as Part of Their Focus

American Society of Ichthyologists and Herpetologists (*Copeia*)
ARCHELON, the Sea Turtle Protection Society of Greece
Asian Turtle Conservation Network (*ATCN Bulletin* and *ATCN Newsletter*)
British Chelonia Group
California Turtle and Tortoise Club
The Caribbean Conservation Corporation (*Velador*)
Chelonian Research Foundation (*Chelonian Conservation and Biology*)
Chicago Turtle Club
Desert Tortoise Council
The Desert Tortoise Preserve Committee, Inc. (*Tortoise Tracks*)
Diamondback Terrapin Working Group
Dutch Turtle & Tortoise Society
Florida Turtle Conservation Trust
Geneva Area Turtle and Tortoise Society
The Georgia Sea Turtle Center (*Caretta Chronicles*)
German Turtle Specialist Group
Gopher Tortoise Council
Gulf Coast Turtle and Tortoise Society
Herpetologists' League (*Herpetologica*)
International Reptile Conservation Foundation (*Iguana Times* and *Iguana*)
International Sea Turtle Society
International Society for the History and Bibliography of Herpetology
    (*Bibliotheca Herpetologica*)

Irish Association of Tortoise Keepers
ISV—International Turtle Association, Austrian Turtle Society
IUCN Marine Turtle Specialist Group (*Marine Turtle Newsletter*)
IUCN Tortoise and Freshwater Turtle Specialist Group (*Turtle and
    Tortoise Newsletter*)
Mid-Atlantic Turtle and Tortoise Society
The Nature Protection Trust of Seychelles
New York Turtle and Tortoise Society
Ontario Turtle and Tortoise Society
Partners in Amphibian and Reptile Conservation (PARC)
Reno Tur-Toise Club
Rio Grande Turtle and Tortoise Club, New Mexico
Sacramento Turtle and Tortoise Club
San Diego Turtle and Tortoise Society
Seattle Turtle and Tortoise Club
Sea Turtle Preservation Society
Sea Turtle Preservation Society Melbourne Beach
Sea Turtle Survival League (*Velador*)
Society for the Study of Amphibians and Reptiles (*Journal of Herpetology;
    Herp Review*)
Society of European Herpetologists (*Amphibia-Reptilia*)
SOPTOM, Le Village des Tortues
Swiss Turtle and Tortoise Interest Group
Tarta Club Italia
Testudinum Societas Cataloniae
Tortoise Group, Las Vegas, Nevada
Tortoise Trust (UK)
Turtle and Tortoise Club of Florida
Turtle and Tortoise Preservation Group
Turtle and Tortoise Society of Charleston
Turtle Survival Alliance
World Chelonian Trust (*World Chelonian Trust Newsletter*)

Magazines and Journals That Specialize in Turtles, Reptiles, or General Herpetology

Amphibia-Reptilia
ATCN Bulletin and *ATCN Newsletter*
Bibliotheca Herpetologica
Caretta Chronicles
Chelonian Conservation and Biology

Copeia
Herpetologica
Herp Review
Iguana
Iguana Times
Journal of Herpetology
Marine Turtle Newsletter
Reptile and Amphibian Magazine
Reptiles Magazine
Tortoise Tracks
Turtle and Tortoise Newsletter
Velador
Vivarium Magazine
World Chelonian Trust Newsletter

Bibliography

Avery, H. W., J. R. Spotila, J. D. Congdon, R. U. Fischer, E. A. Standora, and S. B. Avery. 1993. Roles of diet protein and temperature in the growth and nutritional energetics of juvenile slider turtles, *Trachemys scripta*. *Physiological Zoology* 66:902–925.

Behler, J. L. 1979. *The Audubon Society Field Guide to North American Reptiles and Amphibians*. New York: Alfred A. Knopf.

Bennett, D. H., J. W. Gibbons, and J. C. Franson. 1970. Terrestrial activity in aquatic turtles. *Ecology* 51:738–740.

Bjorndal, K. A., ed. 1981. *Biology and Conservation of Sea Turtles*. Washington, DC: Smithsonian Institution Press.

Bjorndal, K. A., and A. Carr. 1989. Variation in clutch size and egg size in the green turtle nesting population at Tortuguero, Costa Rica. *Herpetologica* 45 (2): 181–189.

Bonin, F., B. Devaux, and A. Dupré. 2006. *Turtles of the World*. Baltimore: Johns Hopkins University Press.

Brown, M. B., G. S. McLaughlin, P. A. Klein, B. C. Crenshaw, I. M. Schumacher, D. R. Brown, and E. R. Jacobson, 1999. Upper respiratory tract disease in the gopher tortoise is caused by *Mycoplasma agassizii*. *Journal of Clinical Microbiology* 37:2262–2269.

Brown, M. B., I. M. Schumacher, P. A. Klein, K. Harris, T. Correll, and E. R. Jacobson. 1994. *Mycoplasma agassizii* causes upper respiratory tract disease in the desert tortoise. *Infection and Immunity* 62:4580–4586.

Buhlmann, K. A. 1995. Habitat use, terrestrial movements, and conservation of the turtle *Deirochelys reticularia* in Virginia. *Journal of Herpetology* 29:173–181.

Buhlmann, K. A., and G. Coffman. 2001. Fire ant predation of turtle nests and implications for the strategy of delayed emergence. *Journal of the Elisha Mitchell Scientific Society* 117:94–100.

Buhlmann, K. A., and J. W. Gibbons. 2001. Terrestrial habitat use by aquatic turtles from a seasonally fluctuating wetland: Implications for wetland conservation boundaries. *Chelonian Conservation and Biology* 4:115–127.

Buhlmann, K. A., T. K. Lynch, J. W. Gibbons, and J. L. Greene. 1995. Prolonged egg retention in the turtle *Deirochelys reticularia* in South Carolina. *Herpetologica* 51:457–462.

Buhlmann, K. A., and T. D. Tuberville. 1998. Use of passive integrated transponder (PIT) tags for marking small freshwater turtles. *Chelonian Conservation and Biology* 3:102–104.

Buhlmann, K. A., T. D. Tuberville, and W. Gibbons. 2008. *Turtles of the Southeast*. Athens: University of Georgia Press.

Burghardt, G. M., B. Ward, and R. Rosscoe. 1996. Problem of reptile play: Environmental enrichment and play behavior in a captive Nile soft-shelled turtle, *Trionyx triunguis*. *Zoo Biology* 15:223–238.

Burke, V. J., and J. W. Gibbons. 1995. Terrestrial buffer zones and wetland conservation: A case study of freshwater turtles in a Carolina bay. *Conservation Biology* 9:1365–1369.

Burke, V. J., J. W. Gibbons, and J. L. Greene. 1993. Prolonged nesting forays by common mud turtles, *Kinosternon subrubrum*. *American Midland Naturalist* 131:190–195.

Burke, V. J., J. L. Greene, and J. W. Gibbons. 1995. The effect of sample size and study duration on metapopulation estimates for slider turtles, *Trachemys scripta*. *Herpetologica* 51:451–456.

Burke, V. J., M. Osentoski, R. D. Nagle, and J. D. Congdon. 1993. Common snapping turtles associated with ant mounds. *Journal of Herpetology* 27:114–115.

Bury, R. B., and J. H. Wolfheim. 1973. Aggression in free-living pond turtles (*Clemmys marmorata*). *BioScience* 23:659–662.

Cann, J. 1998. *Australian Freshwater Turtles*. Singapore: Beamont Publishing.

Carr, A. 1952. *Handbook of Turtles: The Turtles of the United States, Canada, and Baja California*. Ithaca, NY: Cornell University Press.

Carr, A. 1967. *The Sea Turtle: So Excellent a Fishe*. Austin: University of Texas Press.

Clark, D. B., and J. W. Gibbons. 1969. Dietary shift in the turtle *Pseudemys scripta* (Schoepff) from youth to maturity. *Copeia* 1969:704–706.

Conant, R., and J. T. Collins. 1998. *Reptiles and Amphibians: Eastern/Central North America*. Boston: Houghton Mifflin.

Congdon, J. D., G. L. Breitenbach, R. C. van Loben Sels, and D. W. Tinkle. 1987. Reproduction and nesting ecology of snapping turtles (*Chelydra serpentina*) in southeastern Michigan. *Herpetologica* 43:39–54.

Congdon, J. D., A. E. Dunham, and R. C. van Loben Sels. 1993. Delayed sexual maturity and demographics of Blanding's turtles (*Emydoidea blandingii*): Implications for conservation and management of long-lived organisms. *Conservation Biology* 7:826–833.

Congdon, J. D., A. E. Dunham, and R. C. van Loben Sels. 1994. Demographics of common snapping turtles (*Chelydra serpentina*): Implications for conservation and management of long-lived organisms. *American Zoologist* 34:397–408.

Congdon, J. D., and J. W. Gibbons. 1985. Egg components and reproductive characteristics of turtles: Relationships to body size. *Herpetologica* 41:194–205.

Congdon, J. D., and J. W. Gibbons. 1990. The evolution of turtle life histories. In *Life History and Ecology of the Slider Turtle*, edited by J. W. Gibbons, chap. 3. Washington, DC: Smithsonian Institution Press.

Congdon, J. D., and J. W. Gibbons. 1990. Turtle eggs: Their ecology and evolution. In *Life History and Ecology of the Slider Turtle*, edited by J. W. Gibbons, chap. 8. Washington, DC: Smithsonian Institution Press.

Congdon, J. D., and J. W. Gibbons. 1996. Structure and dynamics of a turtle community over two decades. In *Long-Term Studies of Vertebrate Communities*, edited by M. Cody and J. Smallwood, 137–159. San Diego, CA: Academic Press.

Congdon, J. D., J. W. Gibbons, and J. L. Greene. 1983. Parental investment in the chicken turtle (*Deirochelys reticularia*). *Ecology* 64:419–425.

Congdon, J. D., J. L. Greene, and J. W. Gibbons. 1986. Biomass of freshwater turtles: A geographic comparison. *American Midland Naturalist* 115:165–173.

Congdon, J. D., R. D. Nagle, A. E. Dunham, O. M. Kinney, and S. R. Yeomans. 1999. The relationship of body size to survivorship of hatchling snapping turtles (*Chelydra serpentina*): An evaluation of the bigger is better hypothesis. *Oecologia* 121:224–235.

Congdon, J. D., and D. W. Tinkle. 1982. Reproductive energetics of the painted turtle (*Chrysemys picta*). *Herpetologica* 38:228–237.

Congdon, J. D., D. W. Tinkle, G. L. Breitenbach, and R. C. van Loben Sels. 1983. Nesting ecology and hatching success in the turtle *Emydoidea blandingi*. *Herpetologica* 39:417–429.

Congdon, J. D., D. W. Tinkle, and P. C. Rosen. 1983. Egg components and utilization during development in aquatic turtles. *Copeia* 1983:264–268.

Congdon, J. D., and R. C. van Loben Sels. 1991. Growth and body size in Blanding's turtles (*Emydoidea blandingi*): Relationships to production. *Canadian Journal of Zoology* 69:239–245.

Congdon, J. D., and R. C. van Loben Sels. 1993. Relationships of reproductive traits and body size with attainment of sexual maturity and age in Blanding's turtles (*Emydoidea blandingii*). *Journal of Evolutionary Biology* 6:547–557.

Dodd, C. K. 2001. *North American Box Turtles: A Natural History*. Norman: University of Oklahoma Press.

Ernst, C. H., and R. W. Barbour. 1989. *Turtles of the World*. Washington, DC: Smithsonian Institution Press.

Ernst, C. H., and H. F. Hamilton. 1969. Color preferences of some North American turtles. *Journal of Herpetology* 3:176–180.

Ernst, C. H., and J. Lovich. 2009. *Turtles of the United States and Canada*. 2nd ed. Baltimore: Johns Hopkins University Press.

Ernst, C. H., J. Lovich, and R. W. Barbour. 1994. *Turtles of the United States and Canada*. Washington, DC: Smithsonian Institution Press.

Frazer, N. B. 1984. A model for assessing mean age-specific fecundity in sea turtle populations. *Herpetologica* 40 (3): 281–291.

Frazer, N. B., J. W. Gibbons, and J. L. Greene. 1991. Growth, survivorship, and longevity of painted turtles *Chrysemys picta* in a southwestern Michigan marsh. *American Midland Naturalist* 125:245–258.

Frazer, N. B., J. W. Gibbons, and J. L. Greene. 1991. Life history and demography of the common mud turtle *Kinsternon subrubrum* in South Carolina, USA. *Ecology* 72:2218–2231.

Frazer, N. B., J. L. Greene, and J. W. Gibbons. 1993. Temporal variation in growth rate and age at maturity of male painted turtles, *Chrysemys picta*. *American Midland Naturalist* 130:314–324.

Frazer, N. B., and J. I. Richardson. 1985. Annual variation in clutch size and frequency for loggerhead turtles, *Caretta caretta*, nesting at Little Cumberland Island, Georgia, USA. *Herpetologica* 41 (3): 246–251.

Gaffney, E. S. 1990. The comparative osteology of the Triassic turtle *Proganochelys*. *Bulletin of the American Museum of Natural History* 194:1–163.

Gibbons, J. W. 1967. Possible underwater thermoregulation by turtles. *Canadian Journal of Zoology* 45:585.

Gibbons, J. W. 1967. Variation in growth rates in three populations of the painted turtle, *Chrysemys picta*. *Herpetologica* 23:296–303.

Gibbons, J. W. 1970. Reproductive dynamics of a turtle (*Pseudemys scripta*) population in a reservoir receiving heated effluent from a nuclear reactor. *Canadian Journal of Zoology* 48:881–885.

Gibbons, J. W. 1970. Terrestrial activity and the population dynamics of aquatic turtles. *American Midland Naturalist* 83:404–414.

Gibbons, J. W. 1982. Reproductive patterns in freshwater turtles. *Herpetologica* 38 (1): 222–227.

Gibbons, J. W. 1983. Reproductive characteristics and ecology of the mud turtle, *Kinosternon subrubrum* (Lacepede). *Herpetologica* 39:254–271.

Gibbons, J. W. 1986. Movement patterns among turtle populations: Applicability to management of the desert tortoise. *Herpetologica* 42:104–113.

Gibbons, J. W. 1987. Why do turtles live so long? *BioScience* 37:262–269.

Gibbons, J. W. 1988. Turtle population studies. Carolina Biological Supply Company, Burlington, NC, 51:45–48.

Gibbons, J. W. 1990. Sex ratios and their significance among turtle populations. In *Life History and Ecology of the Slider Turtle*, edited by J. W. Gibbons, 171–182. Washington, DC: Smithsonian Institution.

Gibbons, J. W., and J. L. Greene. 1979. X-ray photography: A technique to determine reproductive patterns of freshwater turtles. *Herpetologica* 35:86–89.

Gibbons, J. W., J. L. Greene, and J. D. Congdon. 1983. Drought-related responses of aquatic turtle populations. *Journal of Herpetology* 17:242–246.

Gibbons, J. W., J. L. Greene, and K. K. Patterson. 1982. Variation in reproductive characteristics of aquatic turtles. *Copeia* 1982:776–784.

Gibbons, J. W., G. H. Keaton, J. P. Schubauer, J. L. Greene, D. H. Bennett, J. R. McAuliffe, and R. R. Sharitz. 1979. Unusual population size structure in freshwater turtles on barrier islands. *Georgia Journal of Science* 37:155–159.

Gibbons, J. W., and J. E. Lovich. 1990. Sexual dimorphism in turtles with emphasis on the slider turtle (*Trachemys scripta*). *Herpetological Monographs* 4:1–29.

Gibbons, J. W., and D. H. Nelson. 1978. The evolutionary significance of delayed emergence from the nest by hatchling turtles. *Evolution* 32:297–303.

Gibbons, J. W., and D. W. Tinkle. 1969. Reproductive variation between turtle populations in a single geographic area. *Ecology* 50:340–341.

Iverson, J. B. 1987. Tortoises, not dodos, and the tambalacoque tree. *Journal of Herpetology* 21:229–230.

Iverson, J. B. 1992. *A Revised Checklist with Distribution Maps of the Turtles of the World*. Richmond, IN: John B. Iverson.

Kennett, R., K. Christian, and D. Pritchard. 1993. Underwater nesting by the tropical freshwater turtle, *Chelodina rugosa* (Testudinata: Chelidae). *Australian Journal of Zoology* 41:47–52.

Klemens, M. W. 2000. *Turtle Conservation*. Washington, DC: Smithsonian Institution Press.

Knight, J. L., and R. K. Loraine. 1986. Notes on turtle egg predation by *Lampropeltis getulus* (Linnaeus. Reptilia: Colubridae) on the Savannah River Plant, South Carolina. *Brimleyana* 12:1–4.

Lamb, T., J. W. Bickham, J. W. Gibbons, M. J. Smolen, and S. McDowell. 1990. Genetic damage in a population of slider turtles (*Trachemys scripta*) inhabiting a radioactive reservoir. *Archives of Environmental Contamination and Toxicology* 20:138–142.

Lovich, J. 1988. Aggressive basking behavior in eastern painted turtles (*Chrysemys picta picta*). *Herpetologica* 44:197–202.

Lovich, J. E., S. W. Gotte, C. H. Ernst, J. Harshbarger, and J. W. Gibbons. 1996. Prevalence and histopathology of shell disease in turtles from Lake Blackshear, Georgia. *Journal of Wildlife Diseases* 32 (2): 259–265.

Lovich, J. E., C. J. McCoy, and W. R. Garstka. 1990. The development and significance of melanism in the slider turtle. In *Life History and Ecology of the Slider Turtle*, edited by J. W. Gibbons, 233–256. Washington, DC: Smithsonian Institution Press.

Lutz, P. L., and J. A. Musick, eds. 1997. *The Biology of Sea Turtles*. Boca Raton, FL: CRC Press.

McNamee, G., and L. A. Urrea. 1997. *A World of Turtles: A Literary Celebration*. Boulder, CO: Johnson Printing.

Morreale, S. J., J. W. Gibbons, and J. D. Congdon. 1984. Significance of activity and movement in the yellow-bellied slider turtle (*Pseudemys scripta*). *Canadian Journal of Zoology* 62:1038–1042.

Nicholls, R. E. 1977. *The Book of Turtles*. Philadelphia: Running Press.

Obst, F. J. 1986. *Turtles, Tortoises, and Terrapins*. New York: St. Martin's Press.

Paladino, F. V., M. P. O'Connor, and J. R. Spotila. 1990. Metabolism of leatherback turtles, gigantothermy, and thermoregulation of dinosaurs. *Nature* 344:858–860.

Plotkin, P. T., ed. 2007. *Biology and Conservation of Ridley Sea Turtles*. Baltimore: Johns Hopkins University Press.

Pough, F. H., R. M. Andrews, J. E. Cadle, M. L. Crump, A. H. Savitzky, and K. D. Wells. 2001. *Herpetology*. Upper Saddle River, NJ: Prentice Hall.

Pritchard, P. C. H. 1980. *Encyclopedia of Turtles*. Neptune, NJ: TFH Publications.

Pritchard, P. C. H. 1989. *The Alligator Snapping Turtle: Biology and Conservation*. Milwaukee, WI: Milwaukee Public Museum.

Pritchard, P. C. H. 2006. *The Alligator Snapping Turtle: Biology and Conservation*. Reprint of 1989 edition. Melbourne, FL: Krieger Publishing.

Pritchard, P. C. H., and P. Trebbau. 1984. *The Turtles of Venezuela*. Neptune, NJ: Society for the Study of Amphibians and Reptiles.

Rhodin, A. G. J., P. P. van Dijk, and J. F. Parham. 2008. Turtles of the world: Annotated checklist of taxonomy and synonymy. In *Conservation Biology of Freshwater Turtles and Tortoises: A Compilation Project of the IUCN/SSC Tortoise and Freshwater Turtle Specialist Group*, edited by A. G. J. Rhodin, P. C. H. Pritchard, P. P. van Dijk, R. A. Saumure, K. A. Buhlmann, and J. B. Iverson. Chelonian Research

Monographs No. 5, pp. 000.1-000.38, doi:10.3854/crm.5.000.checklist.v1.2008, www.iucn-tftsg.org/cbftt/.

Ruckdeschel, C., and C. R. Shoop. 2006. *Sea Turtles of the Atlantic and Gulf Coasts of the United States*. Athens: University of Georgia Press.

Scott, D. E., F. W. Whicker, and J. W. Gibbons. 1986. Effect of season on the retention of 137Cs and ^{90}Sr by the yellow-bellied slider turtle (*Pseudemys scripta*). *Canadian Journal of Zoology* 64:2850–2853.

Semlitsch, R. D., and J. W. Gibbons. 1989. Lack of largemouth bass predation on hatchling turtles (*Trachemys scripta*). *Copeia* 1989:1030–1031.

Spotila, J. R. 2004. *Sea Turtles: A Complete Guide to Their Biology, Behavior, and Conservation*. Baltimore: Johns Hopkins University Press.

Steen, D. A., M. J. Aresco, S. G. Beilke, B. W. Compton, E. P. Condon, C. K. Dodd, Jr., H. Forrester, et al. 2006. Relative vulnerability of female turtles to road mortality. *Animal Conservation* 9:269–273.

Steyermark, A. C., M. S. Finkler, and R. J. Brooks, eds. 2008. *Biology of the Snapping Turtle* (*Chelydra serpentina*). Baltimore: Johns Hopkins University Press.

Tuberville, T. D., K. A. Buhlmann, R. K. Bjorkland, and D. Booher. 2005. Ecology of the Jamaican slider turtle (*Trachemys terrapen*), with implications for conservation and management. *Chelonian Conservation and Biology* 4:908–915.

Tuberville, T. D., J. W. Gibbons, and J. L. Greene. 1996. Invasion of new habitats by male freshwater turtles. *Copeia* 1996:713–715.

Tucker, A. D., N. FitzSimmons, and J. W. Gibbons. 1995. Resource partitioning by the estuarine turtle *Malaclemys terrapin*: Trophic, spatial, and temporal foraging constraints. *Herpetologica* 51:167–181.

Tucker, A. D., and N. B. Frazer. 1994. Seasonal variation in clutch size of the turtle *Dermochelys coriacea*. *Journal of Herpetology* 28:102–109.

Tucker, A. D., S. R. Yeomans and J. W. Gibbons. 1997. Shell strength of mud snails (*Ilyanassa obsoleta*) may deter foraging by diamondback terrapins (*Malaclemys terrapin*). *American Midland Naturalist* 138 (1): 224–229.

Vogt, R. C. 2008. *Amazon Turtles*. Manaus, Brazil: Instituto Nacional de Pesquisas da Amazonia.

Wieland, G. R. 1896. Archelon ischyros: A new gigantic cryptodire testudinate from the Fort Pierre Cretaceous of South Dakota. *American Journal of Science*, 4th ser., 2 (12): 399–412.

Wilkinson, L. R., and J. W. Gibbons. 2005. Patterns of reproductive allocation: Clutch and egg size variation in three freshwater turtles. *Copeia* 2005 (4): 868–879.

Wyneken, J., M. H. Godfrey, and V. Bels, eds. 2008. *Biology of Turtles*. Boca Raton, FL: CRC Press.

Yeomans, S. R. 1995. Water-finding in adult turtles: Random search or oriented behaviour? *Animal Behaviour* 49:977–987.

Zim, H. S., and H. M. Smith. 2001. *Reptiles and Amphibians*. New York: St. Martin's Press.

Zug, G. R., L. J. Vitt, and J. P. Caldwell. 2001. *Herpetology: An Introductory Biology of Amphibians and Reptiles*. San Diego, CA: Academic Press.

Index

Page references in italics refer to illustrations. Page numbers followed by *t* refer to tables.

Index 161

Scribner, Kim, 132
seagulls, 51
sea turtles, 2, 5, 36, 100, 129; and biting, 35; eggs of, 57, 59, 65–66; and flippers, 9, 21; geographic distribution of, 6, 7; hatchlings of, 52; and nesting, *62*; and nests, 50–51; and predators, 50–51; shell of, 19; and sleeping, 42. *See also* flatback sea turtle; green sea turtle; hawksbill sea turtle; leatherback sea turtle; loggerhead sea turtle; ridley sea turtles
seed dispersal, by turtles, 54–55
Seidel, Mike, 132
Seigel, Rich, 132
seine, *129*
Semlitsch, Ray, 131
sex determination, 67–68, 102
Sexton, Owen, 133
sexual dimorphism, 13, *14*, 15, *78*
Shaffer, Chuck, 69
Shakespeare, William, "The Phoenix and the Turtle," 120–21
sharks, 51, 106
shell rot, 53
shells, 9, 20, *21*, 134–35; and barnacles, *93*; damage to, 84, *85*, 106; markings on, 27, *73*; morphology of, 30; and predators, 19–20, 27; softening of, 52, 53
side-necked turtle, 2, 5, 43
size, of turtles, 12–16, 135
skunks, 50
sleep, 18, 41–42, *50*
slider turtle, 4–5, 129, *132*; and basking, 31; color of, 29–30; and courtship, 38, 57; Cumberland, 5, 30; and drought, 101–2; eggs of, 58, 59–60, 66; eyes of, 27; and food, 76–78; growth rate of, 70; and learning, 36; and locomotion, 20; and mark-recapture system, 73; and melanism, 28, *29*; and migration, 86, 105; and nesting, 64–65; radioactive contamination of, 103, 104; research on, 130–33; and sexual dimorphism, *14*; shell of, 19, 20; and sleeping, 41; webbed feet of, 21; yellow stripes on, 24–25
slider turtle, red-eared, 5; color of, 30; and food, 76; and melanism, 28; as pet, 82, 129, 133; and predators, 27
slider turtle, yellow-bellied, *3*, 5; in *Charleston Receipts*, *95*; color of, 30; hatchlings of, *47*; and melanism, 28; and play behavior, 37; research on, 131–32
Smith, Michael H., 132
snails, 79–80
snake-necked turtle, 91; northern, 61–62
snakes, 50, 91
snapping turtle, 9, *54*, 129; aggressive behavior of, 34; beak of, 136; and biting, 34, 90, 97; color pattern of, as camouflage, 25; and color vision, 18–19; and combat, 33, 56; consump-

tion of, by humans, 95; and contaminants, 104; defense mechanisms of, 40, 83; eggs of, 57, *66*, 67; and food, 91; and nesting, 64, 65; rescuing, from road, 86–87; sex determination of, 68
social interaction, 31–33
softshell turtle, 5, 9; *Apalone*, 21, 25, 34, 41; Bibron's frog-faced giant, 14–15; consumption of, by humans, 95; eggs of, 57; Florida, *23*, 134; and food, 76; habitat of, 43; New Guinea giant, 14–15; Nile, 37; *Rafetus*, 21; rescuing, from road, 86; and sex chromosomes, 67; Shanghai, 15; shell of, 19, 135; as sit-and-wait predator, 78; size of, 12, 14; smooth, 43; Southeast Asian narrow-headed, 15; spiny, 19, 40, 43; *Trionyx*, 21, 91
Song of the South, 113
Southeast Asian box turtle, 134
Spanish terrapin, 2, 27
species, numbers of, 3–5, 9
speckled padloper tortoise, 16
Spelling, Aaron, 126
spiny softshell turtle, 19, 40, 43
sponges, 95
Spotila, Jim, 17, 131
spots, yellow, 24
spotted turtle, 26, 35, 134
Standora, Ed, 17, 132
Steen, David, 105
Steinbeck, John, *Grapes of Wrath*, 114
streams, 43, 44, 46
stripes, yellow, 24, *25*
Stupendemys geographicus, 10–11
swamps, 2, 7, 43, 44, 49
swimming, 21

taxonomy, 3–5
Teenage Mutant Ninja Turtles, 113–14
teeth, 17
temperature sex determination (TSD), 67–68, 102
terrapins, 2–3
Testudinidae, 2, 5
Thoreau, Henry David, 126
thread trailers, *65*
tidal creeks, 2
titillation, 38, 57
toad-headed turtle, 17
Tong, Haiyan, 10
"tortoise and the hare" algorithm, 115
tortoises, 2, 5, 100; age of, 71; and biting, 34, 35; and color vision, 19; and combat, 56; feet of, 135; and food, 76; in gardens, 90; geographic distribution of, 7; and growth, 70–72; and insulation during winter, 48; longevity of, 74; and mating, 57; shell of, 19–20, 135
tortoiseshell, 105
"Tragedy of the Road" (Blanding), 118–19